Chicken Coops

Chicken Coops

45 Building Ideas for Housing Your Flock

JUDY PANGMAN

Storey Publishing

*The mission of Storey Publishing is to serve our customers by
publishing practical information that encourages
personal independence in harmony with the environment.*

Edited by Sarah Guare and Deborah Balmuth
Art direction and text design by Cynthia McFarland and Kent Lew
Cover design by Kent Lew
Cover photographs by © Getty Images: top row center; © Bruce Goodson: middle row, left;
 © Dennis Harrison-Noonan: top row, left; © Judy Pangman: top row, right; middle row, right
 and bottom.
Back cover illustration by © Elayne Sears
Text production by Jennifer Jepson Smith
Finished coop illustrations by © Elayne Sears
Plan drawings by Brigita Fuhrmann
Indexed by Andrea Chesman

© 2006 by Judy Pangman

Printed in the United States by Versa Press
20 19 18 17 16 15 14 13

Library of Congress Cataloging-in-Publication Data

Pangman, Judy.
  Chicken coops / by Judy Pangman.
    p. cm.
  Includes bibliographical references and index.
  ISBN 978-1-58017-627-9 (pbk. : alk. paper)
  ISBN 978-1-58017-631-6 (hardcover : alk. paper)
  1. Chickens—Housing. I. Title.
SF494.5.P36 2006
636.5'0831—dc22

                                                        2006007688

Acknowledgments

For my parents, John and Lorraine Pangman,
and Frank's mom, Elizabeth Johnson, thank you for everything.
Frank, thanks for going on that hayride twelve years ago.
For Jessie, Randy, Dana, Shannon, Greydon, and Arleigh, and grandkids
Dakota, Ethan, Mikayla, Denver, Austin, Paije, and Haylee:
For Frank and me, the farm is our dream; you are our hearts.
I could not have done this without all of you.

I would also like to thank all of the coop owners for their generous contributions:
Jana Barnhart, Karen Bassler, Herman Beck-Chenoweth, Bryan Boyer, Jennifer Carlson,
Mark and Jodi Clagg, Sharon and Ray Ely, Rena and Gary Georger, Bruce Goodson,
Loren Guernsey, Charles and Linda Gupton, Dennis Harrison-Noonan, Jim and Adele Hayes,
Trish Nickerson, Peter Poirier, Fred Rehl, Joel Salatin, Eric and Melissa Shelley,
Elizabeth Smith, Crane Stavig, Megan and Lynette Terrel, Stephanie Van Parys,
Jennifer Ward, and Eric and Pam Wettering.

CONTENTS

PART 3 Coops for Small Farms

PART 4 Cool Coops

PREFACE

"I WAS GIVEN A FEW CHICKENS," a customer told me at the farmers' market. "You know how that happens — you just end up with them, and then two became six, and the next thing I knew, I was a chicken farmer." I had to laugh because I do know how that happens. We started with a dozen hens a few years ago; now we have more than two hundred laying hens.

Traditionally, most small family farms raised a few chickens in the barnyard. They supplied eggs and meat for the family and neighbors, and surplus was used for barter. During the past 50 years, chicken production followed the exodus from small family farms to factory farms. In this industrial model, hens are packed tightly together in small cages to simplify feeding, watering, and egg collection. Egg production increased dramatically while the hen's life became intolerable.

A growing desire to live in tune with nature and to raise safe and healthy food from humanely treated animals has sparked renewed interest in raising chickens in a natural setting. Rural and urban dwellers alike are seeking ways to connect with the land.

A new phenomenon, the urban chicken, is bringing people together across backyards in many large cities where local ordinances permit residents to raise a few chickens within city limits. As the pastured-poultry movement grows across the country, chickens are helping small family farms make a comeback. From sharing coop designs with aspiring chicken owners to supplying meat, eggs, and compost for their neighbors, chicken owners are making connections, building community, and rebuilding the small-family-farm community.

My husband, Frank Johnson, and I and our young sons Greydon and Arleigh run Sweet Tree Farm, a 200-acre grass-based livestock farm in beautiful upstate New York.

We became interested in chickens when we learned about pastured poultry and discovered how healthy the hens were for our soil. Now, in addition to grass-fed beef and pork, we also produce eggs from pastured laying hens. After six years of raising laying hens, we are still learning. There are even more ways that we could improve our coops and new coop designs that we would love to build.

These pages are filled with a variety of chicken coops, including chick brooders, city chicken coops, pastured laying hen and broiler coops, and other fun and unusual coops. Some of the coops are fancy, and others are made entirely of recycled and salvaged materials. You will be introduced to city chickens and school chickens and will learn a bit about grass farms and pastured-poultry operations.

You may want to build an exact replica of a plan you see in the book, or you may want to adapt a plan to fit your own needs. In order to use the information, you will need to have basic construction knowledge. If you would like more specific building information, read *How to Build Animal Housing* by Carol Ekarius (Storey Publishing, 2004).

Chicken coop owners from across the country generously shared with me their designs and provided tips for using salvaged materials. The features in some of these coops have sparked new ideas (sorry, Frank!) that I would love to implement. I hope that you, too, find lots of ideas in these pages.

1 Providing Shelter: The Basics

THERE IS a lot of information out there, on the Web, in books, and in magazines, about raising chickens, and there are many ways to go about doing it. If you have never raised chickens before, all of this information may be confusing. The most important thing to know is that raising chickens is easy and should be lots of fun!

Chickens have very basic needs: food and water, adequate shelter and space, and predator protection. Chicken owners go to all different lengths to meet those needs, from the simplest to the very elaborate. The beauty of raising chickens is that there is no set formula and no way that you must do it to be successful. You can be as creative and elaborate, or as simple, as you'd like. If you want to nail two old 1-inch × 18-inch boards together into a little A-frame connected by a few furring strips and call it a chicken coop (see the Setting Hen Hut, page 10), you can. Your hen will be just as happy, and her eggs just as beautiful and yummy, as the hen down the street living in the elaborate chicken coop with the gabled roof and painted trim! Chickens can thrive in the country and in city backyards in coops of all shapes and sizes, as long as their basic needs are met.

Chick Brooders

Chicks can be purchased directly from local farm stores and hatcheries. Some stores also take bulk orders for chicks in the spring. A few reliable hatcheries have Web sites and will ship chicks directly to your home or post office (see page 160 for a list of hatcheries). They generally require minimum bulk orders, so if you are buying just a few chicks, you may need to pool your order with chicken-buying friends. If you are purchasing chicks for a laying hen flock, you may want to purchase sexed pullet chicks. They are more expensive than "straight run" chicks (chicks of both sexes), but you will receive mostly female birds. (City ordinances often prevent residents from keeping roosters, and roosters can be difficult to handle!)

We have had great success with chicks that we ordered and received in the mail. We set up the brooder, order the chicks, and have them mailed to our post office box. The hatchery provides an expected delivery date, so we call the postmaster and let him know the date the chicks are expected. When the chicks arrive, the postmaster calls us, and we drop everything and race to the post office. (We are really lucky to live in a very small

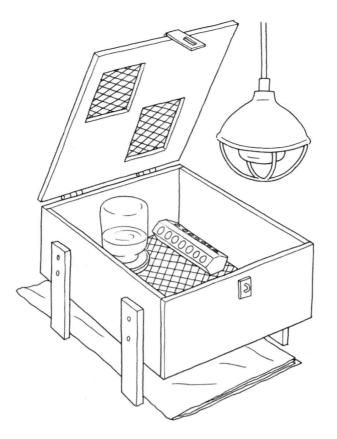

Typical chick brooder with basic equipment: heat source, waterer, feeder, ventilation, and draft and predator protection.

town with a very small post office!) It is quite funny to hear all those peeps coming from that little box behind the counter! Chicks can live without food or water for only a few days after they hatch, so it is crucial to immediately unpack mail-order chicks from their shipping box and give them food and water. As soon as we get the chicks home, we take each one out of the box, place its beak in the waterer for a drink, and release the chick into the brooder.

Brooder Basics

The chick shelter, or chick brooder, should be set up before the hatchlings arrive so that they can be settled in their new home immediately. Those cute little peeps grow up very fast and turn into gangly, flying, feathered "dust balls" that quickly outgrow the chick brooder. If you are building a chicken coop, make sure that it is finished, or very close to finished, before the chicks arrive.

Chicks have a few more requirements than chickens do. The primary difference is that chicks need to be protected from drafts, and

MINIMUM SPACE REQUIREMENTS

BIRDS	AGE	OPEN HOUSING		CONFINED HOUSING		CAGES	
		sq ft/bird	birds/sq m	sq ft/bird	birds/sq m	sq in/bird	sq cm/bird
Heavy	1 day to 1 week	–	–	0.5	20	(Do not house heavy breeds on wire.)	
	1–8 weeks	1.0	10	2.5	4		
	9–15 weeks*	2.0	5	5.0	2		
	21 weeks and up	4.0	3	10.0	1		
Light	1 day to 1 week	–	–	0.5	20	25	160
	1–11 weeks	1.0	10	2.5	4	45	290
	12–20 weeks	2.0	5	5.0	2	60	390
	21 weeks and up	3.0	3	7.5	1.5	75	480
Bantam	1 day to 1 week	–	–	0.3	30	20	130
	1–11 weeks	0.6	15	1.5	7	40	260
	12–20 weeks	1.5	7	3.5	3	55	360
	21 weeks and up	2.0	5	5.0	2	70	450

*or age of slaughter

From *The Chicken Health Handbook,* by Gail Damerow

RATION REQUIREMENTS

TYPE	AGE	RATION	% PROTEIN
Broilers	0–3 weeks	broiler starter	20–24
	3 weeks–butcher	broiler finisher	16–20
Layers	0–6 weeks	pullet starter	18–20
	6–14 weeks	pullet grower	16–18
	14–20 weeks	pullet developer	14–16
	*20+ weeks	layer	16–18
Cocks	maintenance	layer + scratch	9
Breeders	*20+ weeks	breeder	18–20

*Layer or breeder ration should not be fed to pullets until they start laying at 18–20 weeks for Leghorn-type hens or 22–24 weeks for other breeds.
From *Storey's Guide to Raising Chickens*, by Gail Damerow

very young chicks need to be kept warm. You can buy a brooder from a large supplier or from a farm-equipment auction, or you can make your own. Homemade chick brooders can be set up lots of different ways using many different types of equipment and materials. We have raised a few chicks in a plastic storage bin purchased at a local discount store, and now we raise 75 chicks at a time in an old stock-water tank (a large metal watering tub for cows or horses) we found on the farm. Chick brooders can be made from renovated farm sheds or barns, plastic barrel halves, sectioned-off areas of the chicken coop or garage, or other materials and equipment you have available around your home or farm.

Equip the brooder with a heat source and with feed and water containers. Cover the bottom of the coop with bedding to help insulate the chicks and to absorb moisture and droppings. We prefer wood shavings because they smell good and are nice and light for the chicks and for us when we clean the brooder. Other bedding materials include chopped straw, shredded newspaper, peanut hulls, and rice hulls, although these have various problems. For example, some bedding materials absorb too much moisture while others absorb too little, and some materials tend to mat, thus creating a heavy mess to clean out of the brooder.

Feed and Water

Chicks need access to food and clean water at all times. The feeder and waterer need to be kept low enough for the hatchlings to reach. This means that at first, when the chicks are very small, the containers may be right on top of the litter. The chicks will stir the litter, and chances are that much of it will end up in the food and water containers, so plan on cleaning these out a few times during the day. If that is not possible, design the brooder to be large enough so that the food and water containers can be kept at one end with very little litter underneath them.

As the chicks grow taller, the food and water containers should be raised above the litter so they stay clean and fresh longer. We slide wooden blocks under the containers to raise them as needed. The food and water containers can also be hung from the ceiling or the brooder sides to allow the height to be adjusted more easily as the chicks grow. Just make sure not to raise them too high. Chest height to the growing chicks is about right.

A number of poultry-raising books explain the nutrition requirements for chickens, and the table above provides a basic guideline. Our favorite explanation is chapter 3 of *Storey's Guide to Raising Chickens* by Gail Damerow (Storey Publishing, 1995). That is not a shameless plug — we had the book long before we started working with Storey!

You can mix your own feed or buy a pre-mixed ration from a local feed supplier or farm store. We feed our chicks a premixed organic, chick starter prepared by our feed supplier, Cold Springs Farm (see page 159 for contact information). Because organic feed costs about twice as much as conventionally grown feed, we buy it in bulk to cut costs and store it in a bulk-feed storage bin in a corner of our barn. The feed supplier lives only a few miles from our farm, so we arrange for him to truck the feed here and blow it into our feed bin.

The feed is also available in 50-pound bags at our local farm store. In most areas, the local farm store offers commercial chick starter/grower, and many suppliers are beginning to carry organic feed. Commercial chick starter/grower comes in medicated and non-medicated formulas. The key ingredients in most chick-feed mixes are corn for energy and soybean for protein. Organic or locally mixed feeds may have a variety of small grains in addition to the corn, depending on the season and grain availability. Chick starters contain a high amount of protein. As the birds grow, they will need more starch and less protein.

Temperature

Very young chicks need to be kept warm and may require a heat source even during the summer. The heat source can be a brooder heater, heat lamps, or some other type of heater. Start the brooding temperature at 95°F (35°C), and reduce it about 5°F (3°C) each week until the brooder temperature is the same as room temperature.

We hang a heat lamp in one end of the brooder and adjust the height according to the chicks' behavior. If they are loud, are gathering at the outside edges of the brooder, or will not go directly under the lamp, the temperature is too hot. If the chicks are piling up on each other, are loudly peeping, or are huddled directly below the heat lamp, the temperature is too cold. If the chicks are gathering under the lamp with very little fuss and are not piling up or making overly loud peeps, the temperature is just right. A stressed chick sounds much different from a happy chick; you will quickly learn to tell the difference. Happy and content chicks will wander throughout the brooder making quiet little peeps and will sleep side by side without piling up on each other.

Ventilation

Fresh air is good; drafts are fatal. Although it is important that the brooder have adequate ventilation in warm and cold weather, protecting the chicks from drafts is critical.

We make sure that the area in which the chicks live is free of drafts, and we provide ventilation at the top of the brooder. For instance, our stock-tank brooder has high walls that protect the chicks from drafts. In the winter, we cover the tank with old glass storm–door inserts. This creates a greenhouse effect that, with the help of a heat lamp, keeps the brooder toasty warm. It does not provide much airflow, so we slide the window inserts apart so that there is a slight gap at the top of the brooder. This way, fresh air circulates above the chicks, the brooder has adequate ventilation, and the chicks still stay warm and draft-free.

In a larger brooder house, such as a shed or barn, solid walls with windows and vents placed at (human) shoulder level or higher will ensure that the brooder has adequate ventilation and that the chicks are not subjected to drafts.

Predator Protection

Chicks make tasty meals for a variety of critters. Ground predators include raccoons, weasels, foxes, and rats. The best way to protect the chicks from predators is to have solid walls and to cover windows or vent openings with secured small-mesh screening. Check often for holes in the floors and walls and for tears in the screens. Make sure that doors and windows close tightly and that latches cannot be opened by nosy raccoons. If the chick brooder is located inside a building, aerial predators such as hawks and owls are not a problem. Unexpected predators might include household pets or the neighbor's or visitor's dogs. Children must be taught to handle the chicks with care. Avoid attracting predators, such as rats, by storing feed in predator-proof containers and keeping feed-storage areas swept clean.

Chicken Coops

Requirements for chicken shelters vary somewhat depending on whether you are raising laying hens or meat birds. Like chicks, all chickens require adequate shelter, feed, clean water, and predator protection. Laying hens should have some type of nest box, an area in which to lay their eggs, and roosts so they can perch off the ground. Some egg producers also provide additional light during the fall and winter to increase egg production. Meat birds are not provided with roosts because the birds quickly become too heavy to get off the ground.

There are many types of chicken coops. Permanent structures are common. These include stand-alone coops, converted sheds or barns, designated areas within an existing barn or garage, and a variety of coop styles in all shapes and sizes. Mobile coops are growing in popularity, both for laying hens and for meat birds. These are moved around the yard or fields so that the birds have access to fresh grass and bugs and each section of the yard or field receives fertilizer from the chicken droppings. There are two main styles of coops for meat birds raised on pasture. One style keeps the birds confined in the coop directly on the grass; the other allows the birds to range freely inside and outside the shelter within an electric netting fence. What all coops have in common is that they provide for the birds' basic needs.

Feed and Water

Chickens can be fed small amounts of chicken feed at intervals throughout the day, or a feeder can be left out at all times. Feed containers do not need to be fancy. They can be purchased at farm-supply stores or made from buckets, pails, troughs, or other containers that can be easily cleaned. Chickens always need access to fresh water. There are a variety of water containers that work well for chickens, including short plastic or rubber buckets and automatic, vacuum-style, or metal waterers. The food and water containers are kept inside the coop or outside the coop under some sort of cover to protect

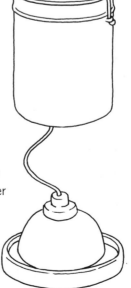

AUTOMATIC HANGING WATERER

The automatic hanging waterer is a very useful system, particularly for mobile coops and pens. Water flows from a hanging pail, through a small hose, to the waterer below. The water should be checked once or twice a day, especially in hot weather. This system is used in Our Farm's A-Frame Henhouse (page 103) and Caretaker Farm's Chicken Prairie Schooner (page 81).

them from the weather. Our mobile coops are high enough that we can hang the feeders and waterers under the coop. We try to locate the food and water away from other high-traffic areas such as the roosts and nest boxes so that they remain free of debris.

Ventilation

Adequate ventilation is important to keep the birds healthy. Vents located near the ceiling on the north and south walls will provide cross ventilation during warm weather. Drafts should be prevented in cold weather. Vents and windows should have covers that can be opened or closed as needed.

Roosts

Chickens instinctively roost in high places at night. Roosts can be made from old or new lumber, tree branches, an old ladder, or other sturdy material. Metal pipes are not the best perches, especially in cold weather, because they can harm the hens' feet. Roosts that are round or have rounded edges are preferable since they are softer on the hens' feet and are easier to grip. Rough or square edges should be filed smooth. Roosts should be checked occasionally to make sure that they are sound and strong. More than one perch can be installed — either side by side with an 18-inch separation or, if space is limited, stair-step style, with at least a 12-inch vertical and horizontal separation. For obvious chicken-dropping reasons, perches should not be placed directly over each other.

Nest Boxes

Laying hens instinctively lay their eggs in dark and protected spaces. Nest boxes are typically installed away from the feeders and waterers. To make egg collection easier, many coop owners design the nest boxes so that they can be accessed from the outside. The nest boxes can be bedded with wood shavings or straw or left bare. Our eggs stay cleaner, and the chickens are less likely to hang around in the nest boxes, when there is no bedding. Others prefer to provide a thick layer of bedding, such as hay, straw, or wood shavings, in the nest boxes to keep the hens comfortable when they lay their eggs. Some coop owners provide a single community nest box in which all of the hens lay their eggs. Other owners provide a series of individual nest boxes.

Nest boxes are available at farm-supply stores and hatcheries. Used metal nest-box units can sometimes be found at farms and farm-equipment auctions, or nest boxes can be built from a variety of materials. A 10-unit nest box is a good model to use when building a nest box from scratch. In this model, two levels of 12-inch-square openings can be accessed from the front or the back, and hinged perches, located across each level, are raised across the opening to prevent the hens from brooding in the nest boxes at night. For just a few hens, an old crate or box filled with shavings or straw will work fine.

Predator Protection

Chicken coops need to be secure from predators in much the same manner as the chick brooder. If the chickens have access to the outside or are housed in mobile coops, larger predators, such as coyotes and foxes, and aerial predators, such as hawks and owls, may become a problem. If the run is small enough, covering the top with chicken wire or mesh will prevent aerial predators from getting to your chickens and help limit the possibility of disease transmitted by wild birds. Electric netting fence is very effective protection against ground predators and is easy to move and set up when the coop is moved. Moving mobile coops often, staying in close proximity to a cattle herd, using livestock guard dogs, avoiding tall trees or poles where owls or hawks can perch, and locating the coops around the farm buildings or where farm activity is busiest will help keep aerial and ground predators at bay.

THINGS TO CONSIDER WHEN DESIGNING YOUR COOP

When you design your coop, consider your property and weather. Think of the worst thing you must design around — hurricane winds, steep hills, below-zero winter weather, and so on. Then build your coop to withstand it. Ask yourself the following questions:

Will the coop you are designing fit through the garden gates, garage doors, and other areas where it will travel? Your chickens are all comfortably ensconced in their beautiful new chicken coop. You hitch up the tractor and proudly tow your creation out to the garden. And your nice new coop won't fit through the garden gate. There it sits. Now you have to either install a new, wider garden gate or rebuild the chicken coop.

Will the coop withstand your weather at its worst? Think wind, rain, lightning, snow, and rising waters. No matter where you raise chickens, it is still farming — and, as any farmer knows, Mother Nature will throw all her nasty little tricks at you when you least expect it and can least afford it. Be prepared.

Can the coop and yard be protected from ground and aerial predators? Nothing is sadder or more frustrating than losing chickens to predators. Chicken wire and portable poultry netting are relatively inexpensive protection against ground predators. Poultry netting with a strong power source (for example, battery, solar, or electric) will effectively fend off large and small critters alike, including coyotes and foxes. Small critters such as rats can get into the coop through even the smallest hole, so check the coop carefully and often for holes, splits, empty knotholes, and signs that rats or weasels are trying to chew through the coop walls or floors. Young chickens often roost at night on the floor near the coop door, and weasels and raccoons can reach beneath the door to kill them. It's not pretty, so make sure that doors are properly sealed and that chicken wire is securely attached at all openings.

Aerial predators such as owls and hawks are difficult to avoid. Small chicken runs can be easily and affordably covered with a predator-proof roof, either with solid roofing materials or with chicken wire. Large pens, especially mobile pastured pens, are more difficult to protect from aerial predators. Some farmers use guard dogs that bond with the flock; others use guard geese or llamas. Aggressive roosters may help some, but they pose additional problems for the safety of the chicken keepers and egg collectors, especially kids. Keeping the mobile pens close to where large animals are grazing may help. Frequently moving the pens keeps predators wondering and makes them less confident in taking the chickens as prey. Keeping the pens near the activity around farm buildings and away from hedgerows and tall structures also helps discourage predators.

Can you easily reach the coop with large food and water supplies or to move it? If you have more than a few chickens, lugging buckets of feed and water gets old really fast. Provide plenty of space for water and feed storage at the coop. If you have lots of chickens, you will want to be able to reach the coop with a wheelbarrow, horse-drawn wagon, four-wheeler, tractor, or truck, depending on your transportation mode. Even a stationary coop in your backyard may have to be moved someday; plan the getaway route in advance and you will not have to take it apart.

Coops for the City

CHICKENS ARE FLOCKING to the city! In some urban areas, local ordinances allow residents to keep a few hens in their backyards. Raising a barnyard animal in the backyard brings new meaning to the change in seasons and the seasonality of food, for adults and children alike. Children learn about life's cycles by caring for hens and collecting eggs. The wonder of cupping a fresh, warm egg in his or her hands is something a child will never forget. Hens are social, fun-loving creatures that get along famously with children. A few hens ranging free in the yard will hang out with the family and quickly adopt the kids as their own.

Your local gardening club may have classes on raising city chickens and building back-yard coops. This would be a good place to start. You can give surplus eggs to neighbors or barter for goods and services, and you can share compost for neighborhood flower beds and gardens. This is the purest model of local food production. Keeping chickens is a great way to reconnect with the land in an urban setting.

When deciding on the type of coop you want to build and where it might be located in your backyard, consider whether you want the chickens to be confined to an area of the yard or to range free. Chickens like to scratch in the dirt and root in gardens and can very quickly damage a perfectly groomed yard. Some chicken owners like to have their hens join them for breakfast on the patio table; others are not thrilled with droppings all over the patio. Hens and children are a good mix, but each should have their own areas for play and rest. You may want a mobile coop and run that can be moved around the yard or a coop that is built in a permanent location.

Whatever style you choose, you can keep building costs down by building a "green" coop using recycled or salvaged materials. Sources for these materials include salvage Web sites, flea markets, garage sales, lumber-yards, salvage shops, farm-equipment sale barns, auctions, and yours or your friends' remodeling projects. Materials to look for include old sheds, windows and doors, old metal nesting boxes, scrap lumber, shingles in broken bundles on sale at the end of the season, plastic for hoop-house covers, used wagon or trailer frames, wagon wheels, chicken wire, and hinges and door latches.

If you live in the city or have a small amount of space in which to build your coop, look at the coops in this section for plans or inspiration.

Setting Hen Hut

HOLDS:
1 hen
1 nest made of
   hay or straw

LOREN GUERNSEY remembers when his grandmother Iantha kept chickens on the family's dairy farm in the 1930s. "She lived in the other half of the farmhouse, and her hens ranged around her side of the yard," Loren explains. "When Grandma had a good setting hen, she built a nest from hay directly on the ground. Then she put the hut over the nest and hen until the chicks were safely hatched and old enough to join the flock."

Sometimes a hen becomes "broody." This means that she instinctively wants to "set" the eggs (keep them warm), gather them under her wings, and protect them from other hens and predators. After the eggs hatch, the chicken cares for the chicks until they are old enough to join the flock. On the Guernsey farm, hens ranged free and found shelter in the farmyard. To keep egg production up and

to have better control of chick production, a few broody hens were separated from the rest of the flock and allowed to set a group of fertilized eggs under the huts. Each Setting Hen Hut held one hen and a nest of fertilized eggs. The hut provided protection from the elements, other nosy hens, and minimally, predators. The hen was typically kept in the hut night and day until shortly after the eggs were hatched.

From scrap lumber, Loren built a replica of the hut he remembers from the 1930s. It has a solid wood back and sides and a slatted front. "I think I got it right," he tells me. "This goes back to when I was just a kid and Grandma was in her 90s. She used to move the hut around the yard after it rained so it stayed clean inside. I remember she used tin pie plates for feed and water because they were low enough to slide under the slats. After the eggs hatched and the chicks got a little bigger, they would run in and out, but they would always go back and tuck in under the hen at night. She did not keep the hen and chicks in the hut too long after that." After the chicks were active enough to run around the yard, the hen would be released with the chicks so that she could offer them some protection from predators and other farm hazards.

Each year during the second week in August, Loren and his wife, Ruth, help run the historical-display building at the annual Cobleskill Sunshine Fair in upstate New York. A replica of an early-1900s kitchen is on display, as well as household antiques, fair memorabilia, and old advertising collectibles. The Setting Hen Hut will join the display of antique farm equipment and machinery. If you get a chance to visit the fair, be sure to say hello to Loren and Ruth. You might find them sitting in the wooden rockers on the front porch of the historical building, dressed in period clothing and happily chatting with fairgoers.

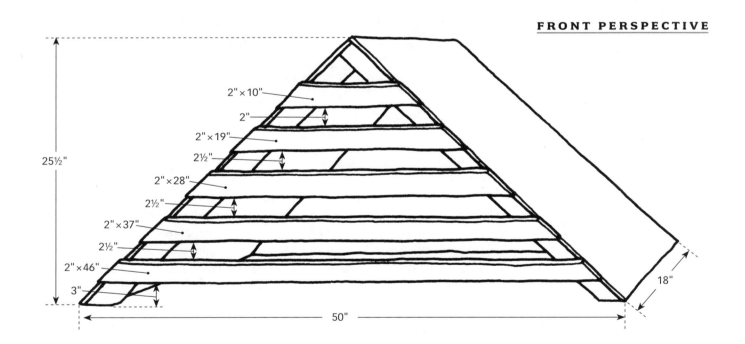

San Miguel Coop

(see color photo on page 154)

"WHEN YOU GO TO SCHOOL for architecture, and someone in your family needs to build something," says Bryan Boyer, "you get drafted to do it for them." So when Bryan's mother wanted fresh eggs and asked him to build her a coop, Bryan and a crew of workers held a coop charette (a community-planning meeting that can last several days) and in four days designed and built this contemporary coop in San Miguel, California.

Bryan explains: "The coop is composed of three main elements, a steel-and-stone monolith, which casts a solid shadow and acts as a deadweight; the actual houses, which sit on a ramp; and a wood-slatted roosting platform, which acts as a shade for the houses below." The coop builders welded the steel monolith in a shop and then trucked their coop to the site. Buried 18 inches in the ground, the monolith is filled with more than a ton of gravel. A steel tray cabled to the monolith supports the wooden roost slats. The henhouses are made from 14-inch ductwork. Bryan originally planned that the chickens would be entirely free-ranged and able to access a nearby tree for shade. When threats from dogs and coyotes made a chicken-wire enclosure a necessity, shade became a priority in the hot California summer. Installing a tarp over a section of the fence solved the problem temporarily; Bryan will eventually design a permanent shade structure.

"My mom has been quite happy with it," says Bryan. "She's happy to have something different, I think, and the chickens seem to enjoy it also."

FLOOR PLAN

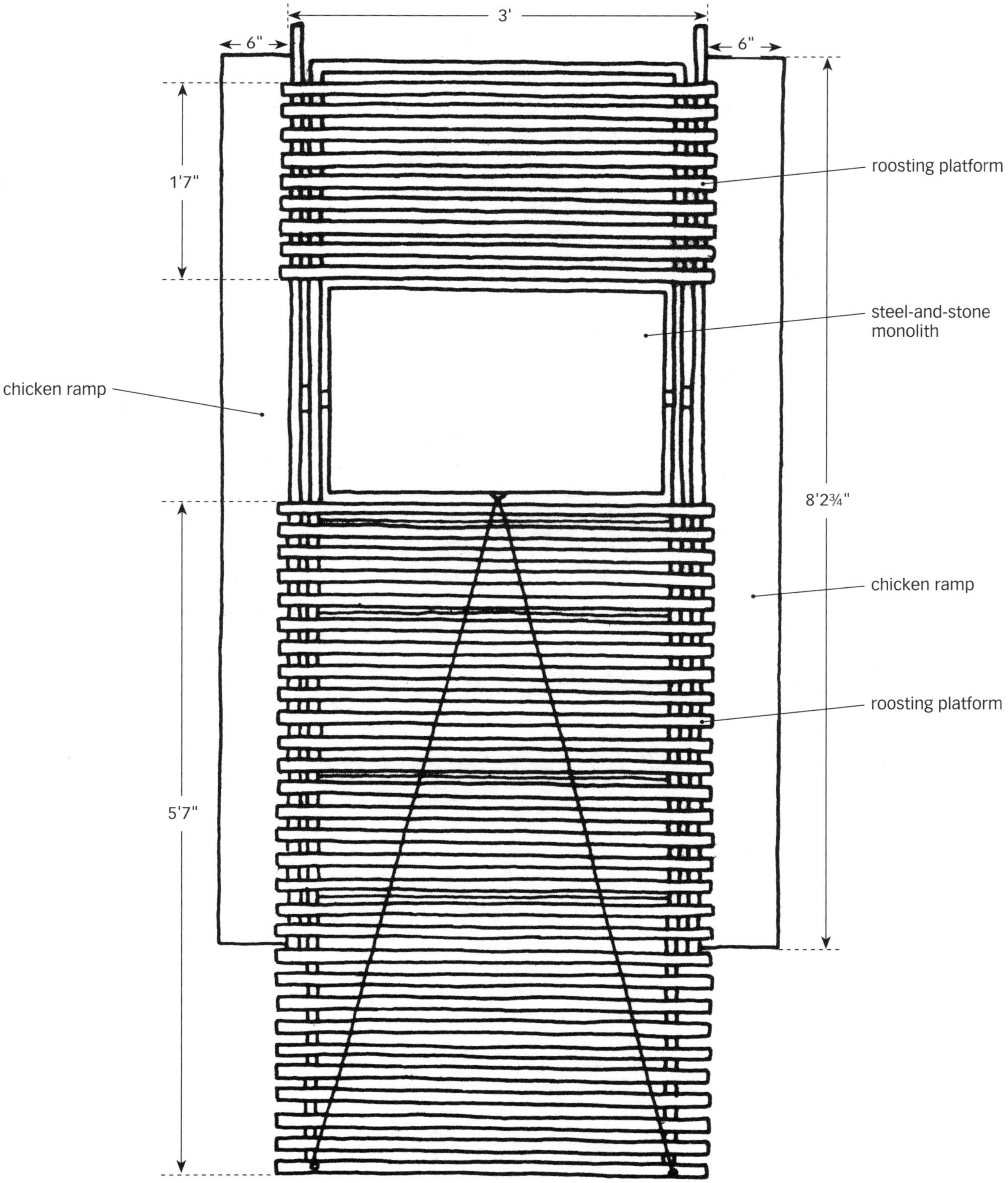
3'
6"
6"
1'7"
roosting platform
steel-and-stone monolith
chicken ramp
8'2¾"
chicken ramp
roosting platform
5'7"

San Miguel Coop (continued)

SIDE ELEVATION

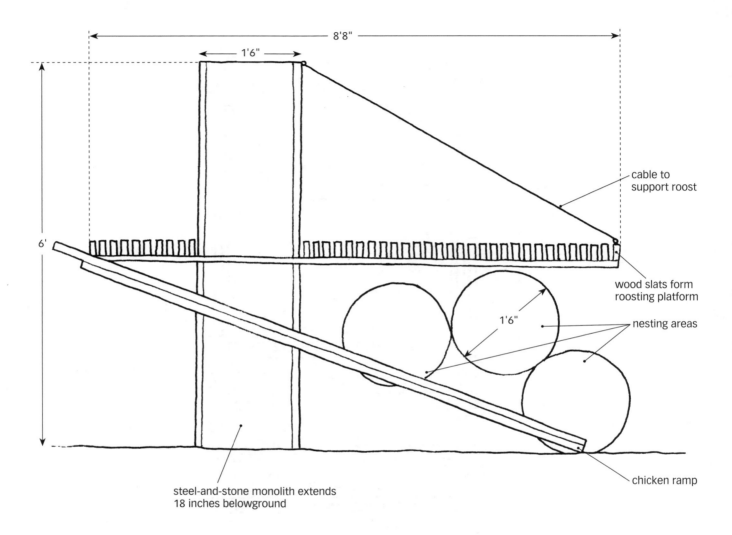

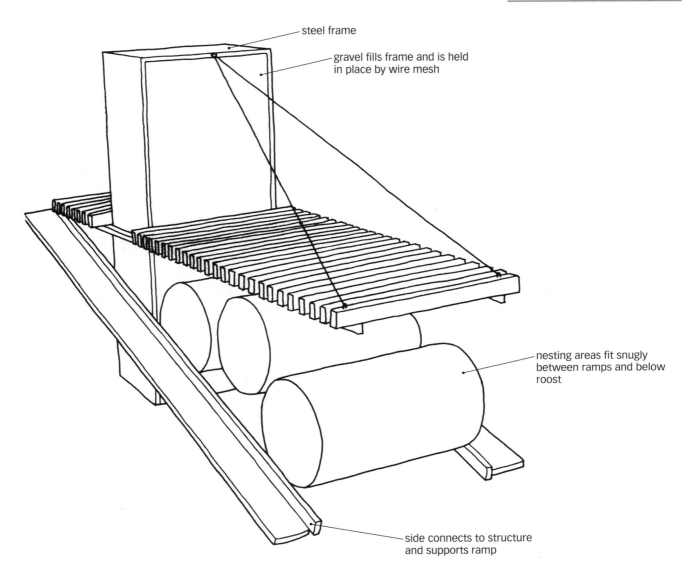

steel frame

gravel fills frame and is held in place by wire mesh

nesting areas fit snugly between ramps and below roost

side connects to structure and supports ramp

Bo & Friends Coop

IF YOU VISIT the Troy Waterfront Farmers' Market on a Saturday morning in upstate New York, be sure to spend a few minutes visiting with the resident celebrity. Bo greets visitors, poses for pictures with them, and entertains the kids with his antics. He is a media favorite and is often featured in newspaper and magazine articles. Everyone knows Bo, so if you can't find him right away, someone will point him out to you. Bo has lots to crow about these days!

Bo is Our Farm CSA's mascot, a spangled Old English bantam rooster that lives at Jenn Ward's vegetable-and-poultry farm. Bo spends his market day perched on Our Farm CSA's sandwich-board sign at the entrance to the farmers' market. Back at the farm, while the other chickens live and work in the various pasture coops in the gardens, Bo and the farm's poultry celebrities-in-training live in a cushy backyard coop near the house. Bo is joined by fancy bantam hens and roosters, as well as by a few rare dominique chickens that fellow farmer Erika Marczak shows at local fairs and brings to the Glens Falls Farmers' Market.

Bo's coop is an old 20-bushel apple box. One of the original crate walls was removed and replaced with a chicken-wire combination wall and door that opens into a yard made from a metal puppy crate. The squash crate is covered with a slanted shingle roof, and the pen is covered with a sheet of plywood. The birds can be closed in the pen at night; during the day, they are allowed to roam where they will around the backyard.

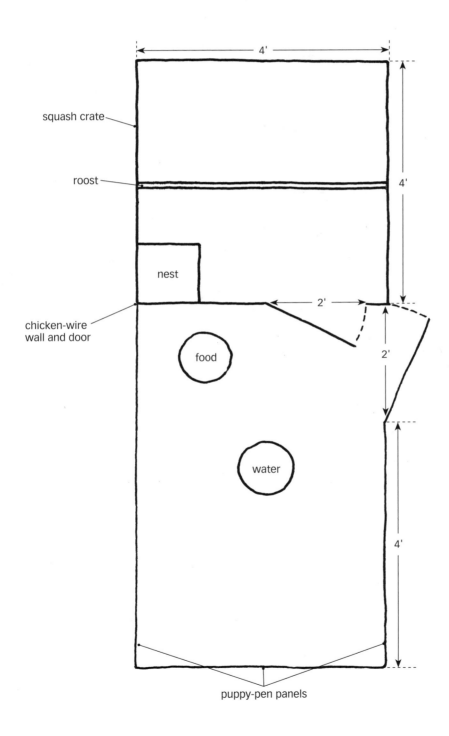

FLOOR PLAN

squash crate

roost

nest

chicken-wire
wall and door

food

water

puppy-pen panels

Bo & Friends Coop (continued)

FRONT ELEVATION

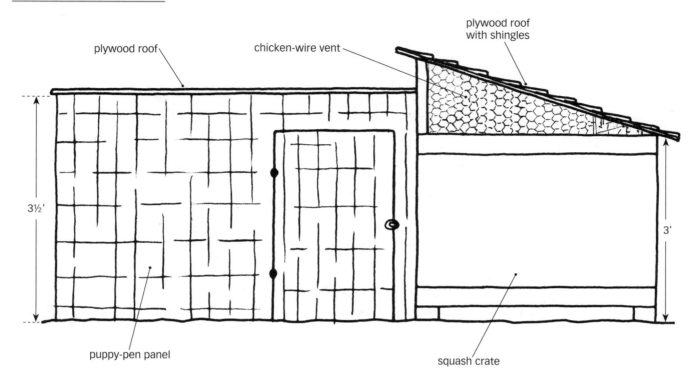

plywood roof

chicken-wire vent

plywood roof
with shingles

3½'

3'

puppy-pen panel

squash crate

Cézanne's Garden Coop

(see color photo on page 151)

HOLDS:
4 hens
1 rooster
3 nest boxes
2 roosts (plus
the rafters!)

"I HAVE WANTED to have chickens for years, but being in the city, I did not think it was possible," says Jana Barnhart. "I talked to my nearest neighbor to make sure she was okay with it before I started. When I am out walking in the morning, I can hear a rooster a couple of blocks over, so it seems I am not the only one with chickens!"

Jana developed her chicken-coop design after chatting with other chicken owners on a popular chicken-enthusiast Web site called Backyard Chickens (see page 162 for Web address). "I was low on money but had big dreams. I gathered together scraps of wood from here and there and found a picture on the Internet of what I had in mind. My husband, who was initially very unsupportive, decided to get involved, and the framework went up. I was so excited! It looks like a child's playhouse so as not to draw too much unwanted attention." Jana's Web site (see page 162 for Web address) includes pages devoted to her chickens and coop, pond, garden, family, and business.

This quirky coop has plenty of room for the hens and rooster. Designed to blend with the garden, the coop has an uneven roofline (it dips toward the center) and cedar-shake roof, painted trim around the lopsided windows, and hand-painted accents. The three-unit nesting box juts out from the side of the coop, and a hinged roof on the box allows for access from the outside. A waterer, a feeder, and two roosts are located inside the coop. Doors on the side open to a chicken run that winds around the back edge of the garden and is enclosed with a chicken-wire hooped roof. A heat lamp is turned on during the winter as needed to keep the chickens warm.

"I have one Buff Silkie rooster named Maurice," says Jana. "There is one Black Silkie hen and three 'Easter egger' [Araucana] hens. The EE hens just started laying, and I am getting some really pretty blue-green eggs."

Cézanne's Garden Coop (continued)

FLOOR PLAN

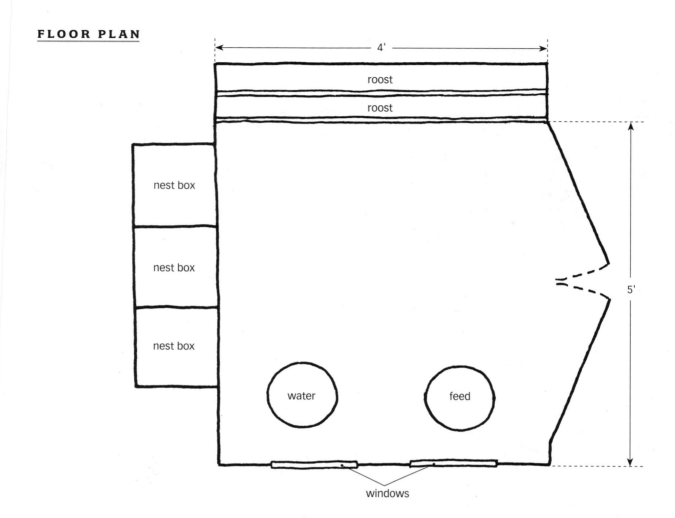

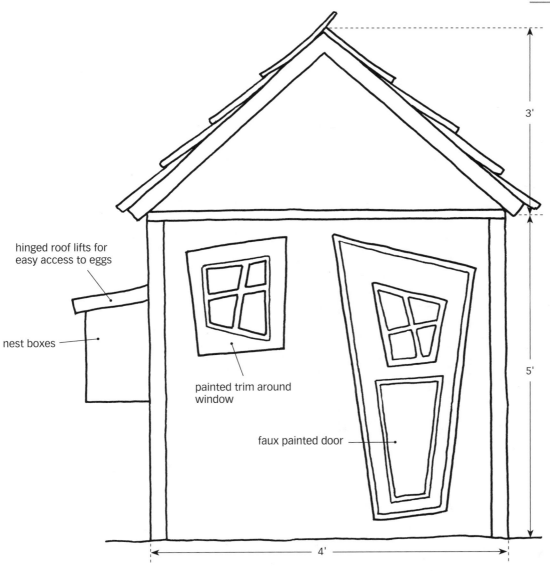

hinged roof lifts for
easy access to eggs

nest boxes

painted trim around
window

faux painted door

3'

5'

4'

Cézanne's Garden Coop (continued)

SIDE ELEVATION

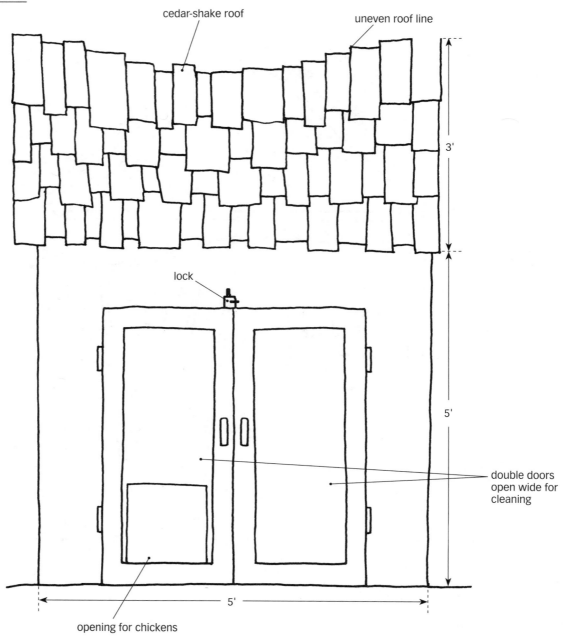

cedar-shake roof

uneven roof line

lock

3'

5'

double doors
open wide for
cleaning

5'

opening for chickens

Sun Coop

(see color photo on page 154)

HOLDS:
6 hens
1 rooster
1 nest box
3 roosts

"WE BROUGHT DAD six chicks to distract him from Mom's illness. We were halfway home after seeing the chicks at a farm store when I made my husband, John, turn the car around so we could go back and buy the chicks," says Kathy Ritter. "Dad kept the chicks in the house until the coop was ready. Bringing the chicks home was one of the best things we did for Dad!"

Kathy's dad, Fred Rehl, had recently purchased a new garden shed that he installed at the edge of the garden near the house. After the chicks came, he separated a portion of the shed with a chicken-wire wall so that half remained a garden shed and the other became a chicken coop. Fred installed a hen-sized door so the hens could have access to a small outdoor pen.

"The supplier considers these sheds to be upscale," says Fred. "So when the company representative called me to see how I liked our new shed, imagine his surprise when I told him that we had converted the shed into a chicken coop. There was silence on the line for a few minutes while he absorbed that news!"

The hens and rooster spend the summer in and around the Sun Coop and roam free in the garden when Fred is home. As soon as the weather turns cold in the fall, Fred loads them up and takes them to a nearby farm, where they live out the rest of their days. Fred was worried about the quality of the new home. Then, when one of the chickens had breathing problems, the farmer rigged up a chicken-sized oxygen tent. The chicken stayed in the tent and recovered! So now Fred doesn't worry about the chickens anymore. Next spring, he will start again with six new chicks for the Sun Coop.

FLOOR PLAN

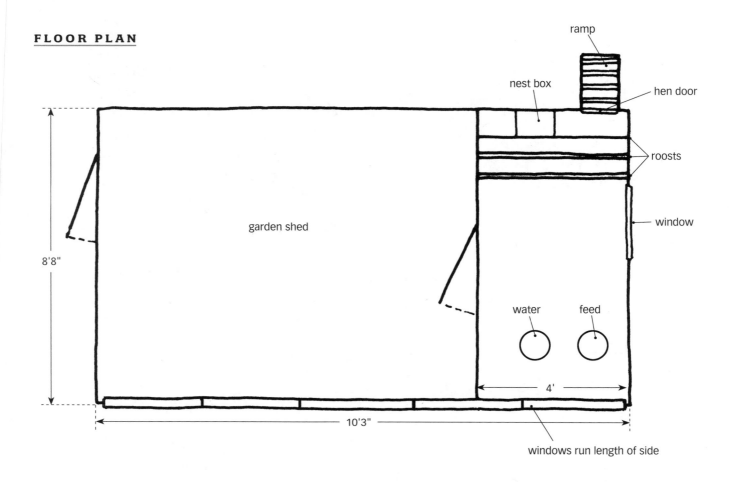

ramp

nest box

hen door

roosts

window

garden shed

8'8"

water

feed

4'

10'3"

windows run length of side

PUNKIN HOLLER GANG

In a tradition that began more than 50 years ago, a bunch of Fred's friends (and their kids and grandkids) get together every summer in a group called the Punkin Holler Gang. They scrounge around for old bathtubs, outhouses, water tanks, chicken coops, farm equipment, and whatever else they can find and put this collection of stuff on old trucks, wagons, and lawn mowers. The jalopies are decorated with misspelled signs and old junk, and crates are added for whatever barnyard animal the gang can convince to ride along. The gang members dress up in hillbilly costumes, and the whole lot of people, critters, and jalopies join parades throughout New York State. That first year, the Rehls taught their favorite hen, Jesse, to walk on a leash. They painted her toenails red, and she joined the Punkin Holler Gang. Needless to say, Jesse was the hit of the parades!

FRONT ELEVATION

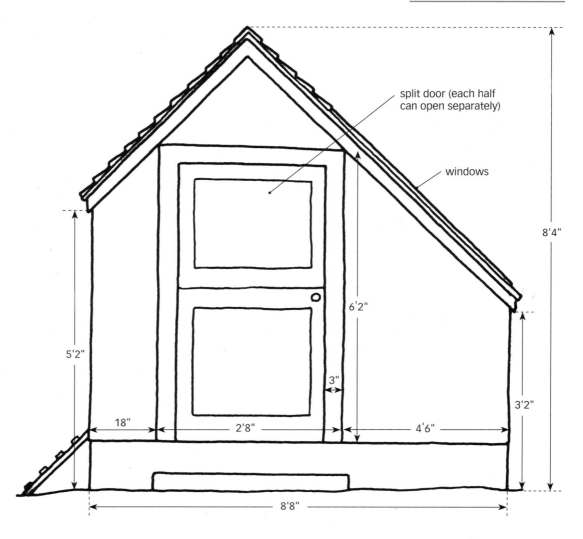

split door (each half can open separately)

windows

8'4"

6'2"

5'2"

3"

3'2"

18"

2'8"

4'6"

8'8"

Sun Coop (continued)

SIDE ELEVATION

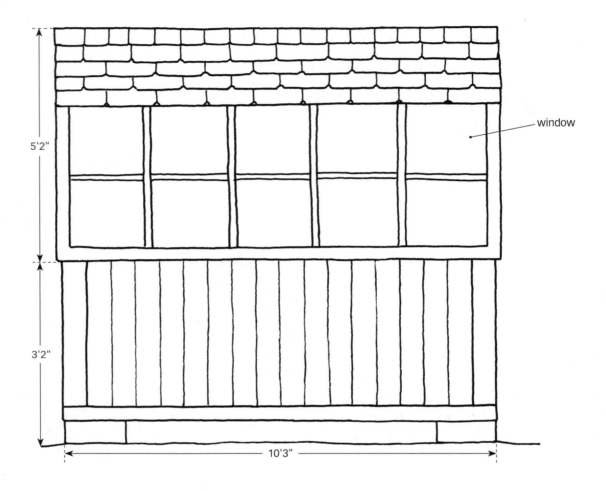

window

5'2"

3'2"

10'3"

Megan's Coop

A FEW YEARS AGO, nine-year-old Megan Terrell's grandfather, Papa Clarence, helped her show chickens at the Greene County Youth Fair near their home in upstate New York. From then on, Megan was hooked on raising chickens. She soon convinced her parents to build her a chicken coop. The entire family pitched in and built Megan's Coop over the course of the next summer. Megan's dad, Keith, designed the coop and cut all of the boards for the frame and floor. Megan and her mom, Lynette, nailed the floorboards and siding. Megan's six-year-old sister, Amanda, helped carry boards and held them in place while the others pounded nails. The white picket fence was painted by Megan's 96-year-old great-grandmother, Henrietta Smigel.

The coop is a 10-foot × 10-foot structure separated into two sections: one 4 feet × 10 feet, and the other 6 feet × 10 feet. The hens live in the small part, and the larger part is used for food storage. The two coop sections are separated by a wall. The lower half of the wall is made of plywood. The upper half, which runs up to the ceiling, consists of 2-inch × 3-inch sections of block wire. A storm door is located in the middle of the wall. The coop is designed so that Megan can handle the chickens all by herself. She can come into the coop and close the main door to the outside. This way, she can feed the birds, clean the coop, collect the eggs, and spend time with the chickens without worrying that they might escape.

The inside walls are lined with rigid-board insulation and covered with sheets of plywood. The ceiling, lined with rigid-board insulation but not covered in plywood, has a single lightbulb fixture in the center. During the winter, the light is left on and provides enough heat to keep the hens comfortable. The coop is built on pressure-treated 4×4s placed on concrete blocks. The floor is made of hardwood floorboards. The chicken section has an eight-hole nest box secured to the back wall. A 4-foot section of thick dowel is secured to the wall next to the nest box for a roost. On the other side of the nest box, a 2-foot-long wooden trough feeder is placed on the floor. A large coffee can is used for water. A hinged and latched chicken door located on the bottom of the coop provides access to the fenced side yard.

The side yard is surrounded by a picket fence and covered with block wire to keep the hens in and the predators out. The storage section holds feed barrels and other supplies. It also contains a small crate for a Cochin hen and rooster that were recently given to Megan by friends. Megan's rooster is dominant, so the new hen and rooster are kept separated from the rest to avoid fighting.

Most of the materials used to build the coop were recycled from their previous uses.

FLOOR PLAN

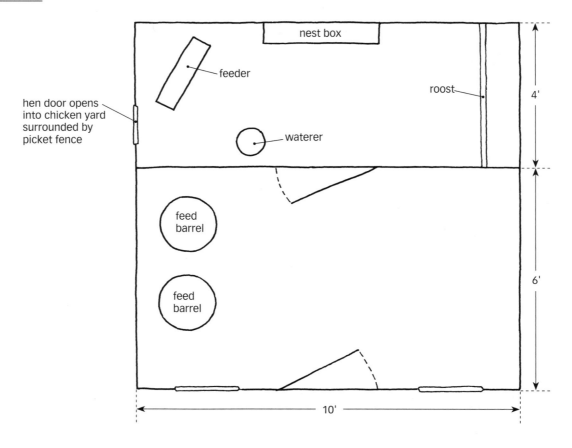

The outside of the coop was finished with cedar siding contributed by friends who were replacing the siding on their home. The old siding had been painted gray, so Keith turned it over to reveal the natural cedar. The coop's windows were salvaged from a dump, and the doors and insulation came from neighbors. The eight-hole nest box and the outside picket fence were given to Megan by her great-aunt Henrietta. The metal roofing was purchased new and was the costliest part of the coop.

The flower boxes were built by Megan and her dad. They share coop duties: Keith handles the morning chores so that Megan can have time to get ready for school. As soon as she gets off the bus at the end of the school day, Megan does the afternoon chores. She shovels out the pen when it needs cleaning, collects the eggs, feeds the chickens, and checks to make sure that they have enough water. Keith sells all of the eggs at work. Megan spends a lot of time with her chickens. One hen in particular, Aurora, seems to have adopted Megan as her own. While Megan was showing us her coop, Aurora wandered over and Megan picked her up. The next thing we knew, Aurora jumped to Megan's shoulder and perched there while we continued our tour of the coop!

FRONT ELEVATION

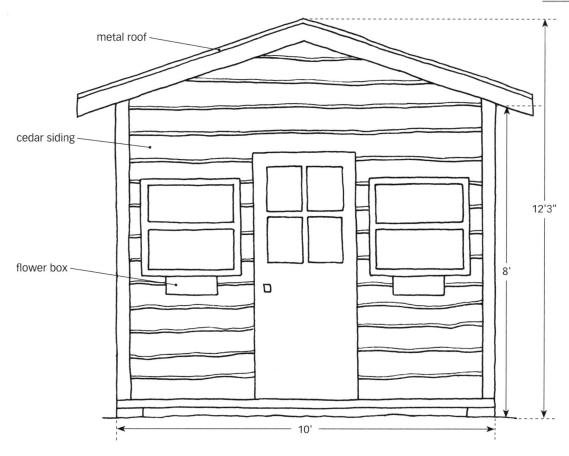

metal roof

cedar siding

flower box

12'3"

8'

10'

Cottonwood School Coop

HOLDS:
10 hens
6 nest boxes
6 roosts

AT THE COTTONWOOD SCHOOL, a Montessori school in Corrales, New Mexico, hens are happily ensconced in a wood-framed adobe coop built by the schoolchildren (with a little help from the adults). The hens are an integral part of the school's permaculture curriculum as a hands-on gardening activity. "The chickens started out as a mini-ecosystem model as part of the environmental-science program. Children feed the hens their lunch scraps; the hens eat the bugs from the garden and fertilize the ground; the ground grows carrots, which we eat and then compost or feed to the chickens — who lay the eggs, which we also eat and cook with, send to the homeless shelter, and sell to buy more chicken food!" explains Trish Nickerson. "It is a great way to see actual environmental and life cycles such as the egg-to-chick cycle."

The Cottonwood School chicken coop was designed to have a "face" when seen from the back. It has two top windows for eyes and a low window at the bottom as a mouth. The children remud the coop once a year or as needed for repairs. The wire enclosure surrounding the coop keeps the coyotes out. The coop was originally covered with wattle and daub, an African method of weaving and mudding, but is now covered primarily with adobe each spring.

Caring for the hens reinforces three primary permaculture ethics: take care of the earth, take care of the people, and reinvest the surplus. Nine classes of children ages 2 through 12 work in the garden in their weekly environmental-science class. During the one-hour class, the children care for the chickens, compost, recycle, gather water, mulch, propagate seeds in the greenhouse, garden, and harvest. "It's just another way of demonstrating actual life cycles while allowing the children to take responsibility for the plants and animals," Trish explains. "The indirect lesson is that we are all connected in the web of life — fully interdependent — so that when one person makes a decision to care (or not care), it affects everybody."

The children's chores are extensive. They collect and weigh the eggs each day, store them in a refrigerator near the garden, sell the eggs to parents, and donate the excess to the homeless shelter. The children also check the food and water bins, change the straw bedding, plant shade plants around the outside of the coop, and mud the coop as needed for repairs.

FRONT ELEVATION

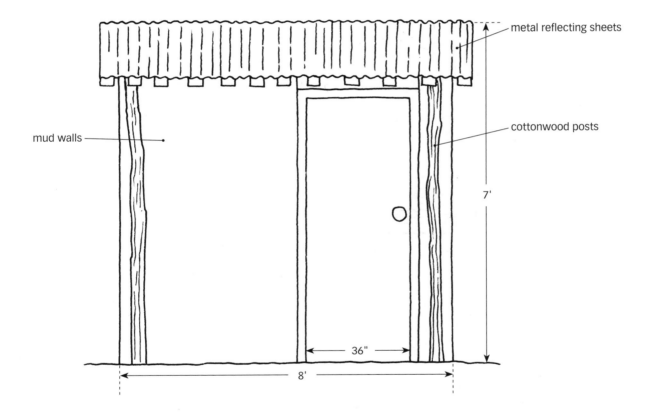

metal reflecting sheets

cottonwood posts

mud walls

7'

36"

8'

Cottonwood School Coop (continued)

SIDE ELEVATION

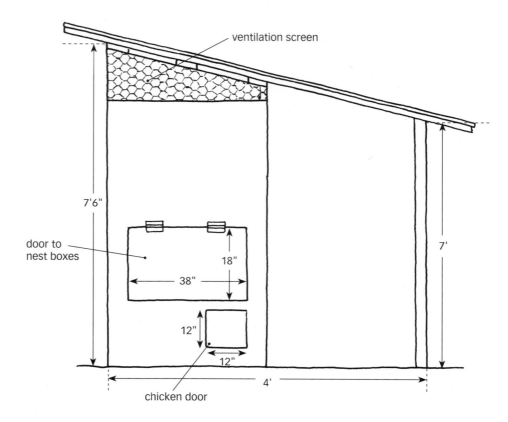

ventilation screen

7'6"

door to
nest boxes

18"

38"

12"

12"

chicken door

7'

4'

BACK ELEVATION

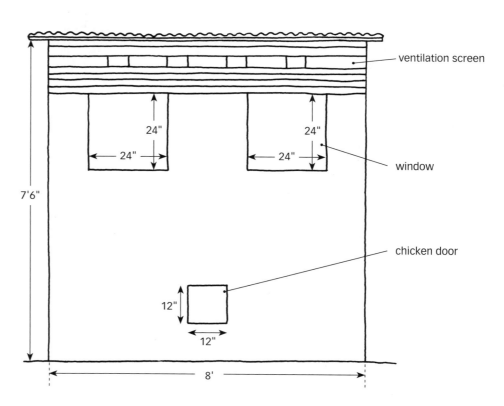

ventilation screen

24"

24"

24"

24"

window

7'6"

chicken door

12"

12"

8'

Oakhurst Community Garden Coop

(see color photo on page 151)

(see color photo on page 151)

HOLDS:
5 hens
2 nest boxes
2 roosts

WHEN CHILDREN vandalized a garden in a Decatur, Georgia, neighborhood in 1997, neighbors joined forces and asked the kids to become caretakers of the garden. The neighbors mentored the kids as they restored the flower beds and shrubs and painted the garden fence. The kids soon went from destroying their world to caring for it, and they loved it! From this early success, and with a local resident's gift of a long-term lease on a half-acre lot, the Oakhurst Community Garden Project was born. The gardens are within walking distance of four schools, and the project's many gardening and outdoor programs teach local students to notice and care for the environment. In addition to the gardens, the project hosts a henhouse; a mature woodlot and wetland meadow; and habitats for bees, rabbits, and other native wildlife.

The Oakhurst Community Garden Project chicken coop is in the back part of the garden, near the meadow and stream. The coop has a small footprint, but it stands tall. It is elevated 16 inches above the ground on posts and has a pitched roof. Sliding windows on both sides allow for cross ventilation. A small (4-foot × 8-foot) enclosed yard was built with the original coop. This yard has a door and lid for easy access and is surrounded by chicken wire, which has been driven 1 foot into the ground to protect against burrowing predators. An outer fence was built to give the chickens more room to wander around. That fence was built using 5-foot welded wire fencing and is held in place with tall garden stakes. Chicken food is stored outside of the coop inside metal trash cans that sit on top of a pallet.

▶ The roof is 5' × 9' with 2×4 framing, ½" plywood, and three-tab shingles.

▶ The siding is TL-11 with 1× trim.

▶ The windows are covered with ¼" hardware cloth.

▶ The coop is supported by 4×4 posts on deck blocks.

▶ The 2'6" door is made from TL-11 siding with 1× trim and has strap hinges.

THE OAKHURST COMMUNITY GARDEN PROJECT

The Oakhurst Community Garden Project teaches environmental awareness to diverse local students through hands-on gardening and outdoor education programs. Project leaders are dedicated to empowering young people to become active members of the community by engaging in projects that address real needs. Respect for the earth and each other underlies all their efforts. By teaching about wellness, teamwork, and conservation, they hope to empower youth to take charge of their own health, as well as that of the environment. More information about the Oakhurst Community Garden Project can be found on their Web site (see page 162 for Web address).

FLOOR PLAN

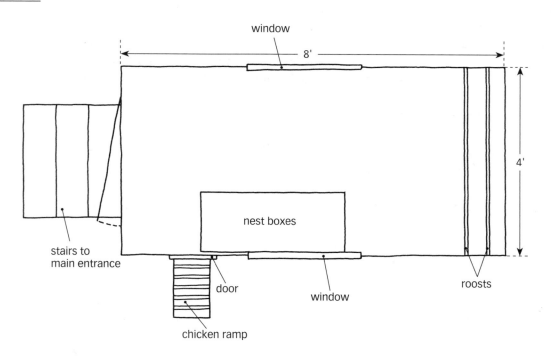

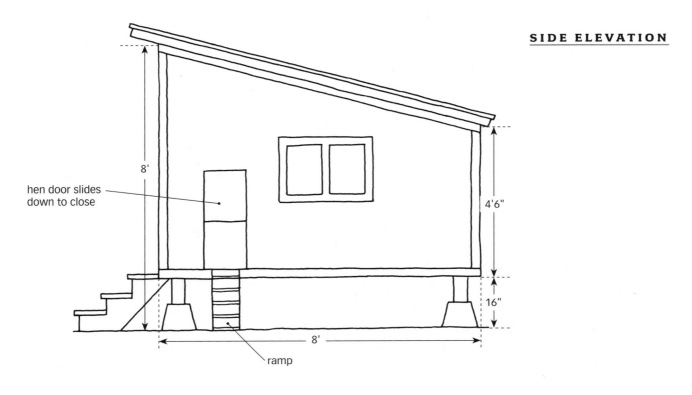

8'

hen door slides
down to close

4'6"

16"

8'

ramp

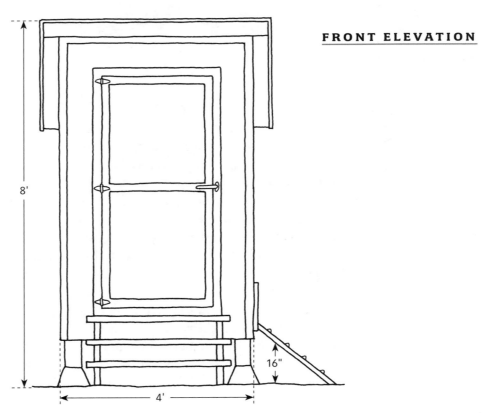

8'

16"

4'

Cordwood Chicken House

(see color photo on page 152)

HOLDS:
16 hens
1 rooster
6 nest boxes
8 roosts

PAM WETTERING KNEW her chickens would not be warm enough in a regular coop when winter temperatures drop to between −20 and −40°F (−29 and −40°C) at her Tower, Minnesota, property. So Pam carved a semicircular chicken house with 10-inch-thick cordwood walls into the base of a hill. "We chose the cordwood design because we had originally planned on building a cordwood home," says Pam. "We had wanted to make the cordwood home semicircular, set into the hill, just like the chicken house. We figured the chicken house would be perfect practice in building with the cordwood theme. And if we messed up, it would be with the chicken house, not ours!"

The temperature in the chicken house stays between 15 and 40°F (−9 and 4°C) during the winter. For about one month in the winter, outdoor temperatures range from −20 to −40°F (−29 to −40°C). On those days, Pam turns on a small propane heater that is hung on the wall of the chicken coop. Constructing the coop took about two months. Pam did most of the work herself, with some help from her husband, Eric, a semitrailer driver who is away five days out of seven.

The base of the hill was excavated enough to set a concrete floor tight against the hill. With no running water available at the time, Pam and Eric hauled water from the lake a mile away and mixed a total of 50 bags of concrete by hand in a wheelbarrow. This formed a 4-inch insulated concrete slab for the coop floor. Because the coop would be recessed into the hill, Pam was concerned that the hill might eventually push the finished structure off the concrete pad. To prevent this from happening, she installed rebar rods in the concrete at the front end on each side of the pad, set concrete blocks over the rebars, and poured concrete into the concrete-block holes. The 10-inch-thick walls were placed behind these anchor blocks. The walls are birch and aspen log ends, or cordwood, cut from trees on the Wetterings' 40 acres. Pam laid a line of mortar on each end of the logs and placed sawdust in between them for insulation (see sidebar at right). She repeated this process in layers until the walls were roughly 6 feet tall in the front and sloped to 5½ feet in the back, so that rain and snowmelt could run off the roof. The corners are 12-inch lengths of 4×4s stacked in an alternating pattern.

The roof frame and sills are 4×4 lumber covered by plywood, sealant, and roll roofing. The roof is covered with 2-inch rigid insulation, a layer of black plastic sheeting, and 4 inches of soil. Pam originally planted grass and wildflowers on the roof, but because the chickens preferred using the soil for their dust baths, the plants never took root. Drain tile was placed around the sides and back of the coop. Pam covered the ends of the wood facing the hill with two layers of mortar, roll roofing, rigid insulation, and black plastic sheeting before shoveling earth back around the walls to smooth out the slope of the hill. It worked — the walls have not leaked. Windows facing the south and east provide light and ventilation.

The interior of the house has an electrical outlet and a light. The portable propane heater is used only during the few coldest weeks of the winter. When the heater is on, Pam pulls out a loose log end to ensure adequate ventilation for the hens. The nesting box has three levels. The lowest is for the broody hen and her chicks, the second level is for the rest of the flock, and the third level opens from the top and is used for feed storage. The chickens access the chicken-wire-enclosed deck through a small chicken-sized door in front. They can get out to the fenced-in yard at the side through a second chicken door. The human door gives them access to the rest of the yard and to the forest.

"My husband, Eric, daughter, Sierra (age 5), and myself now can sit outside and thoroughly enjoy watching our chickens peck, scratch, take dust baths, and just do goofy chicken things!" says Pam.

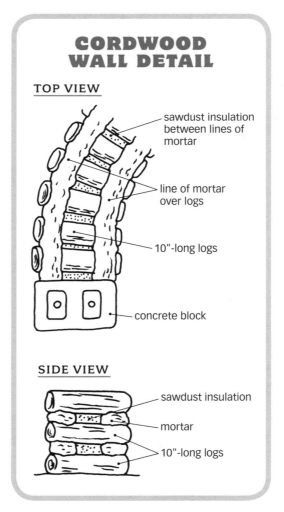

CORDWOOD WALL DETAIL

TOP VIEW

sawdust insulation between lines of mortar

line of mortar over logs

10"-long logs

concrete block

SIDE VIEW

sawdust insulation

mortar

10"-long logs

Cordwood Chicken House (continued)

FLOOR PLAN

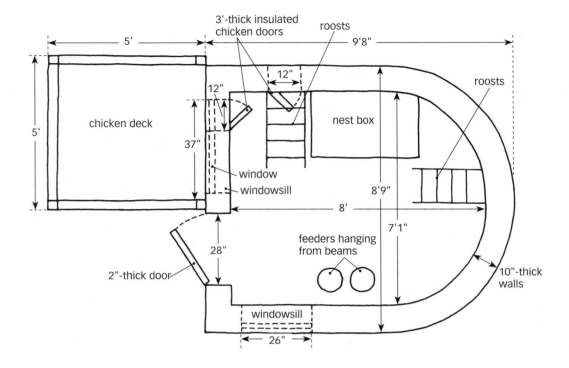

BEAM LAYOUT

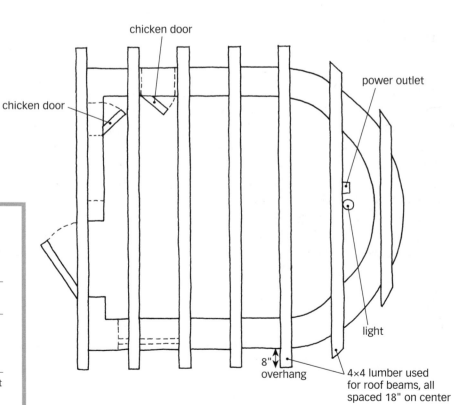

CONSTRUCTION NOTES

▶ The feeders have been suspended from the beams. You could use any tough rope or metal hook to do this.

▶ The chicken doors have been insulated with 3-inch-thick rigid insulation.

▶ The lower-level nest box has removable dividers that make three separate nest boxes or, when removed, create a brooding area.

▶ Two sets of roosts look like ladders that are set at an angle against the wall.

2×8 lumber, set 2" out from structure

7½"

8"

10"

37"

28"

6'1"

38"

6'8½"

5'

chicken door
to deck

12"

12"

8'9"

NEST BOX UNFINISHED

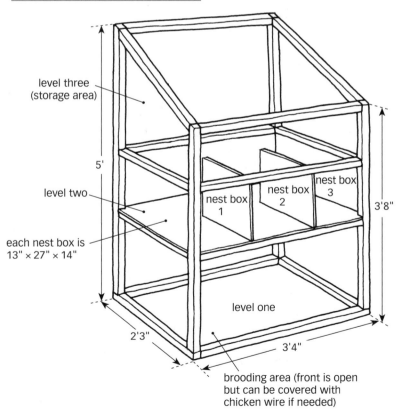

level three
(storage area)

level two

each nest box is
13" × 27" × 14"

5'

3'8"

nest box
1

nest box
2

nest box
3

level one

2'3"

3'4"

brooding area (front is open
but can be covered with
chicken wire if needed)

NEST BOX FINISHED

flip-up cover

flip-up door

chicken entry
into nest
boxes and
brooding area

5"

Little Red Henhouse

(see color photo on page 153)

HENS AT the Little Red Henhouse can be viewed 12 hours a day, 7 days a week on the live "chicken cam" (see page 162 for Web address). "Our interest in chickens stems from my wife's work at a local sanctuary, where she discovered that chickens are unique, affectionate animals, each with its own personality and preferences," explains coop owner and builder Peter Poirier. "My allergies had so far precluded us from having any of the more conventional pets, so chickens seemed a good match, and we certainly haven't been disappointed."

This popular coop on the City Chickens Tour in Seattle is based on a design from an article titled "The Movable Coop" in the September/October 2001 issue of *Organic*

Gardening. The basic design was changed with the addition of a removable extension run and a foundation. The coop is constructed of fir, plywood, and some cedar. Usable untreated wood was difficult to find at local construction sites, so, for the most part, new materials were used. The plastic corrugated roof lets in light. The screening is ½-inch-square wire mesh. The foundation is a wood frame made with 2×4 boards that have been treated with raw linseed oil and set on the ground. Interior surfaces are treated with canola oil. Exterior surfaces are painted with latex paint because it is nontoxic when dry.

The coop consists of two levels. The lower level is surrounded by wire mesh, has no floor, and opens into the attached run. The feeder is suspended by a hook in the lower level. One side of the second level has a ladder to the ground level and two nest boxes that are accessible from a small drop-down door on the side. The other side has a single roost and the waterer. One side of the roof is hinged just under the ridgeline so that it lifts up to allow access to the second level. The hens are kept inside the coop during the mild Seattle winters, and no additional winterization to the coop is needed.

"We raised our three hens from chicks born this past March, and since then we have become very attached to them," says Peter. "We plan on keeping them their entire lives and just using and sharing the eggs with our neighbors."

FRONT PERSPECTIVE

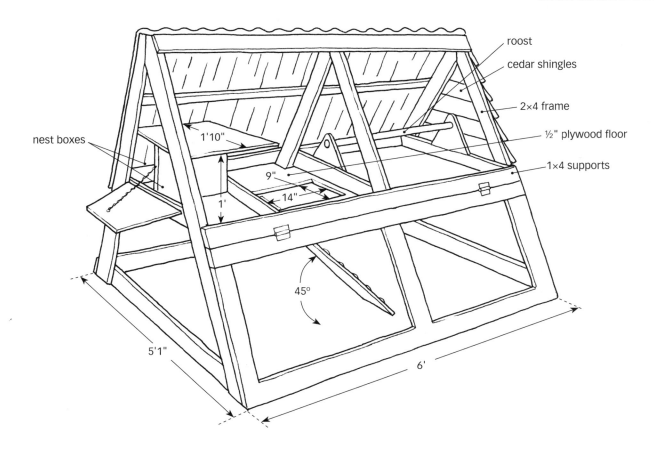

nest boxes · roost · cedar shingles · 2×4 frame · ½" plywood floor · 1×4 supports · 1'10" · 9" · 14" · 1' · 45° · 5'1" · 6'

Little Red Henhouse (continued)

SIDE ELEVATION

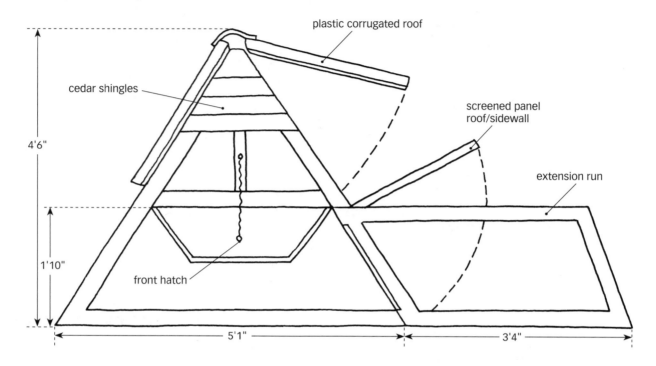

plastic corrugated roof

cedar shingles

screened panel roof/sidewall

extension run

4'6"

1'10"

front hatch

5'1"

3'4"

SEATTLE TILTH CITY CHICKEN TOUR

IN MANY CITIES that allow homeowners to keep a few hens in their backyards, community gardening groups such as Seattle Tilth are offering courses that teach potential chicken owners how to house and raise city chickens. Organized in 1978, Seattle Tilth is one such organization that is cultivating a sustainable city, one backyard at a time. With more than one thousand members, Seattle Tilth connects the city's residents with the land through more than one hundred organic-gardening classes and special events. As these connections are made, participants connect with each other to build a sense of place and community. City chicken owners learn from each other, trade coop-design ideas, give guided tours of their coops to the public, and network with other city chicken groups.

"Seattle Tilth is the most urban of the Tilth chapters, and our base has always been in the Wallingford neighborhood, where our current demonstration garden was a part of the master plan for transforming the Good Shepherd Center into a community center," says Karen Holt Luetjen, executive director. "We added a children's garden in 1988, and another demonstration garden in Seattle's south end, which was conceived in 1995 and dedicated in 2004."

The Seattle Tilth mission statement is to "inspire and educate people to garden organically, conserve natural resources, and support local food systems in order to cultivate a healthy urban environment and community."

In addition to a comprehensive organic-gardening program, recent classes offered by the association include Fall Salad Gardening, Cultivating Plant Communities in Your Garden, Heirloom Gardening — Seed Saving, and Putting the Gardens to Bed.

Special events include a harvest fair and an edible-plant sale. Many of the association's organic-gardening classes and special events are held in Seattle Tilth's demonstration gardens.

For chicken enthusiasts, there are the City Chickens classes. In the City Chickens 101 class, students of all ages learn everything they need to know about raising chickens in their own backyard. City Chickens 201 is the advanced class.

On a Saturday in July, residents join the annual City Chickens Tour and spend the day visiting backyard chicken coops throughout Seattle. Many of the coops on the tour were built after their owners attended the City Chickens 101 class. These coops may house just a few chickens, but their owners are serious chicken folk. The coop designs are as imaginative as their owners and reflect a fierce pride both in having backyard chickens and in being included on the tour.

Hi-Rise Coop

(see color photo on page 154)

HOLDS:
3 hens
1 nest box
2 roosts

THREE LUCKY Buff Orpington hens are living the high life in the Hi-Rise Coop in Jennifer Carlson's Seattle backyard. The coop was built in 2000 as a demonstration project for an exhibit at the Northwest Flower and Garden Show, held in February at the Washington Convention Center. Along with other students in the University of Washington Landscape Architecture program, Jennifer created several components of the coop, a wheelchair-accessible garden, and a vegetable garden. "Over 80,000 people attend the F&G Show every year," Jennifer says, "so it was a great opportunity to show people how to have a productive, vibrant garden in an urban setting."

Including chicken coops in the Northwest Flower and Garden Show is a natural fit because chickens can be an integral part of home gardening. Chickens are great insect and weed controllers, especially between plantings and between growing seasons. Adding chicken manure to the garden bed provides needed organics and nutrients.

Jennifer's interest in chickens began in 1979 when she purchased 12 Buff Orpington day-old chicks as part of a large mail order placed by Seattle Tilth members. (Since then, Seattle has passed an ordinance that restricts the number of chickens to three per household.) The coop design evolved after Jennifer moved a number of times and had to leave behind coops that either were too large to move or were attached to other structures. Jennifer teaches Seattle Tilth's class on chicken-coop building, and plans with complete specs can be purchased from Seattle Tilth (see page 162 for Web address).

The colorful Hi-Rise Coop consists of three separate modules that can easily be moved by two people to another part of the garden or onto a truck for transport. The main module has a nest box on the second level and feed-storage space on the lower level. A ladder connects the nest-box area with the next module. The second module's roof and solid walls provide protection for winter living and keep the feed and water dry. The third module also has a roof but has wire walls for comfortable summer living. An optional "chicken tractor" (a 2-foot × 2-foot × 8-foot wire run) can be attached to the last module or can be moved around the yard to allow the birds to graze on grass. The wire run and nest box contain 6 inches of bedding. The modules sit on concrete pavers on top of 6 inches of ⅝-inch crushed rock. Concrete pavers are placed in front of the coop for easy cleaning and for wheelchair access.

The coop is built using 2×2 framework with ⅜-inch CDX plywood sheathing and cedar siding. Old vinyl flooring salvaged from a kitchen remodel serves as the floor in the feed-storage box to ensure easy cleaning. The vinyl flooring also acts as a vapor barrier to keep the feed dry. For additional cost savings, the cedar shingles were purchased as broken bundles at the end of the summer. The nesting box can be accessed by lifting the roof, or it can be built with a hinged front panel so that the box is accessible to kids and seniors. The coop is easily cleaned with a small rake and trowel, and Jennifer washes it with disinfectant twice a year.

FRONT ELEVATION OF FIRST TWO MODULES

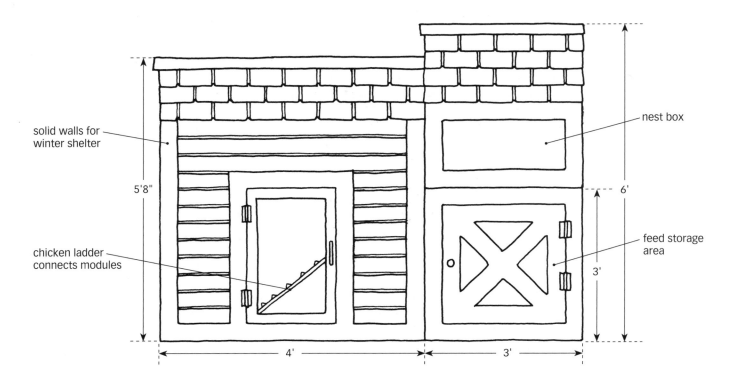

solid walls for winter shelter

nest box

5'8"

6'

feed storage area

chicken ladder connects modules

3'

4'

3'

Hi-Rise Coop (continued)

CONSTRUCTION NOTES

▶ The modules can be configured any way you'd like. Additional modules can be added to the wire run for extra run space.

▶ Half-inch wire is best for the screen netting.

▶ A copper ridge along the top of the roof was added to better shed water in heavy rains.

▶ The storage box contains feed, bedding, oyster shells, and cracked corn.

FRONT ELEVATION OF THIRD MODULE

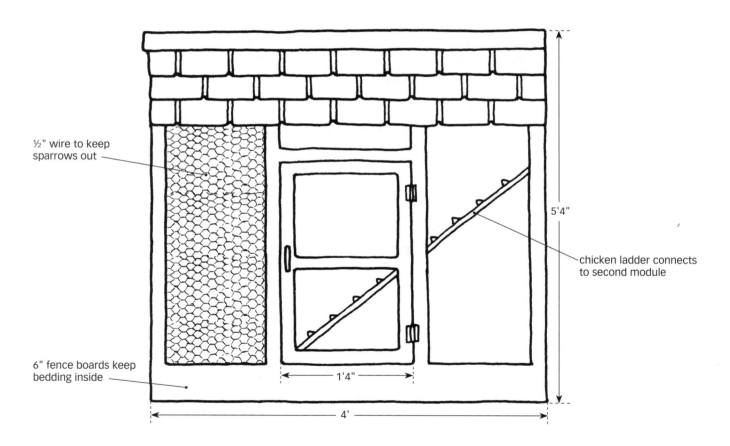

½" wire to keep sparrows out

6" fence boards keep bedding inside

chicken ladder connects to second module

5'4"

1'4"

4'

Poulet Chalet

(see color photo on page 151)

ITS CLEAN LINES and transparent roof make the Poulet Chalet a Seattle Tilth City Chickens Tour favorite, and one that everyone wants to build for their backyard. "My objective was to make a coop that was portable, secure from predators, and easy to maintain and clean," coop owner Bruce Goodson explains. "So far, I have been very happy that the design has met these objectives." Lightweight and easily picked up by two people gripping the boards on the side, Poulet Chalet is completely portable. The coop is entirely enclosed with wood and screening to make it secure from predators. One side of the roof is hinged at the top, and on the opposite side, half of one of the bottom screens is a hinged door to allow for plenty of access to clean and maintain the coop.

Poulet Chalet houses three chickens comfortably and has plenty of room inside for a light, a feeder, and a waterer. The second level has a sturdy tree-limb roost and a nest box that can be accessed from the outside via a small door located on the outside of the coop. The roost and nest box are located in the upper level of the coop, and chickens use a ramp to reach the lower ground level, where the feeder and waterer are suspended. The ramp is pulled up by a cord to secure the chickens from predators overnight. The chicks can access the yard through a chicken door on the side. The chickens stay warm and comfortable in Seattle's mild winters. In areas with severe winter weather, the coop could be moved into the garage or barn for added protection against the elements.

Poulet Chalet (continued)

FLOOR PLAN: GROUND LEVEL

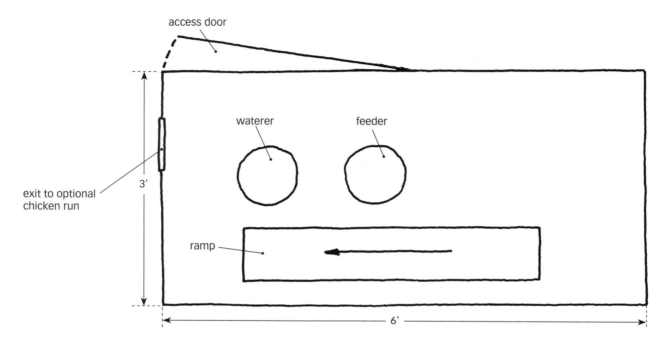

FLOOR PLAN: UPPER LEVEL

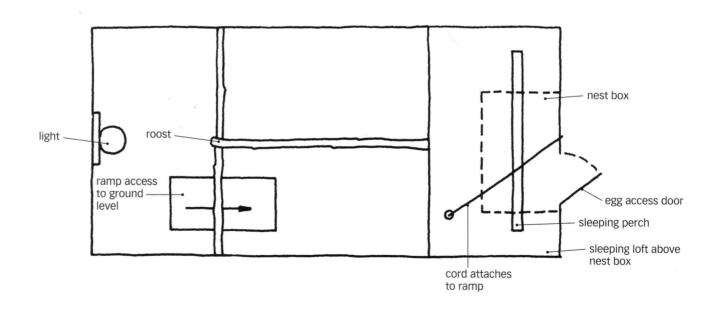

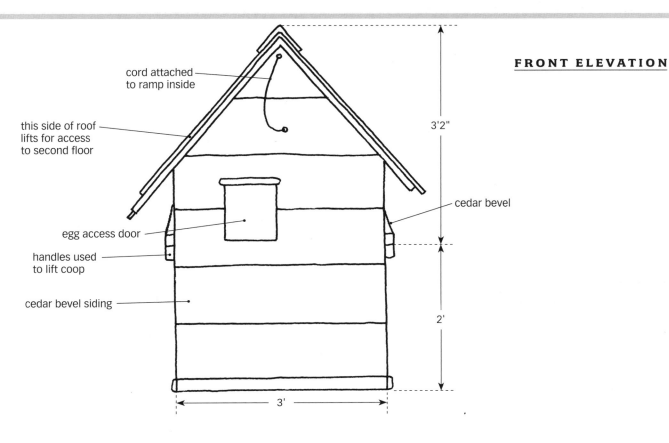

cord attached
to ramp inside

this side of roof
lifts for access
to second floor

3'2"

cedar bevel

egg access door

handles used
to lift coop

cedar bevel siding

2'

3'

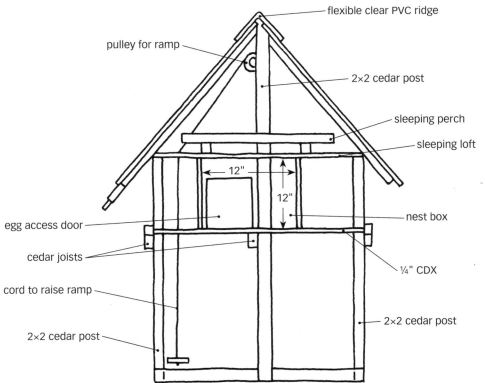

flexible clear PVC ridge

CROSS SECTION

pulley for ramp

2×2 cedar post

sleeping perch

sleeping loft

12"

12"

egg access door

nest box

cedar joists

¼" CDX

cord to raise ramp

2×2 cedar post

2×2 cedar post

Poulet Chalet (continued)

SIDE ELEVATION

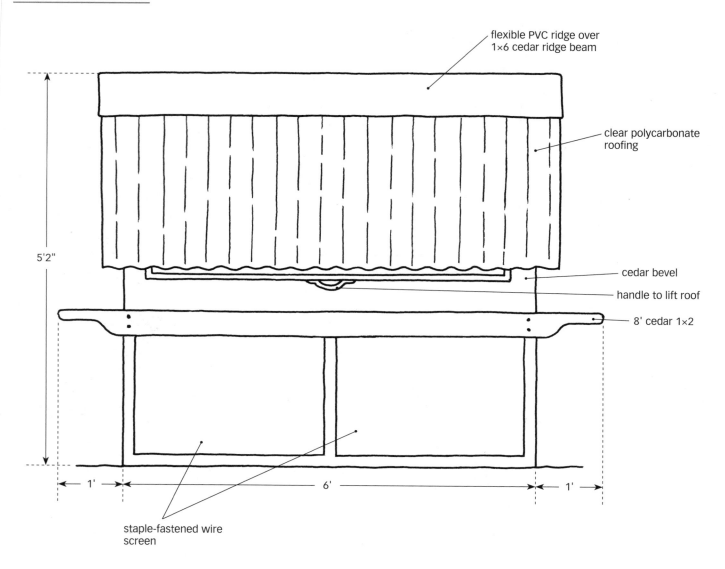

flexible PVC ridge over
1×6 cedar ridge beam

clear polycarbonate
roofing

cedar bevel

handle to lift roof

8' cedar 1×2

5'2"

1'

6'

1'

staple-fastened wire
screen

The Cat House

(see color photo on page 152)

THE CAT HOUSE is a real work of art in the Seattle Tilth City Chickens Tour. The Georger family built the coop to house four hens in their Seattle backyard. The coop is built on a 4-foot × 6-foot pallet and made entirely of salvaged materials. The nesting boxes (total of four) are accessed from a drop-down recycled door across the back. Two roosts are located on the second tier inside the coop. An old door with large windowpanes lets in light and opens for cleaning.

The hens wander into the attached run via small doors at the bottom of the coop, and hen doors are located on the outside so the hens can occasionally roam the yard. The main entrance to the chicken run can be closed and latched at night. A lean-to off the side of the coop and open space underneath the roosts and nest boxes provide shelter on rainy days. The coop, with its roof mosaic and hen mural painted by John Weik, is the focal point of the yard.

The Cat House (continued)

FRONT PERSPECTIVE (UNFINISHED)

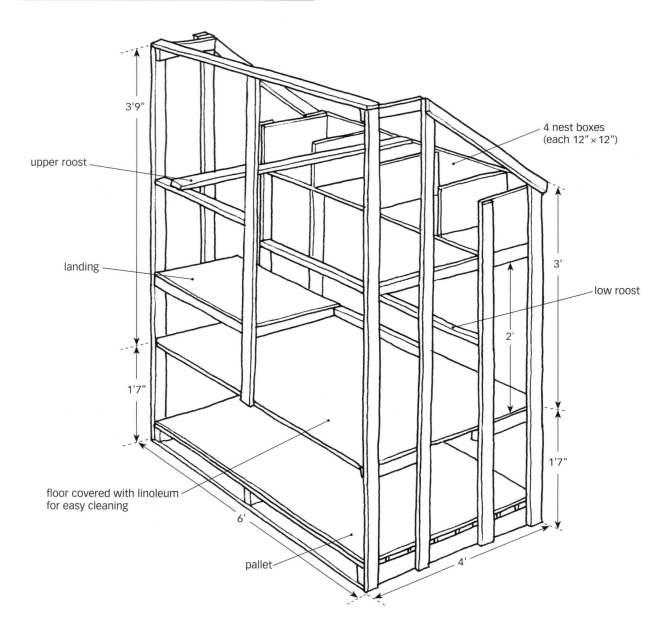

3'9"

upper roost

landing

1'7"

floor covered with linoleum
for easy cleaning

6'

pallet

4 nest boxes
(each 12" × 12")

low roost

3'

2'

1'7"

4'

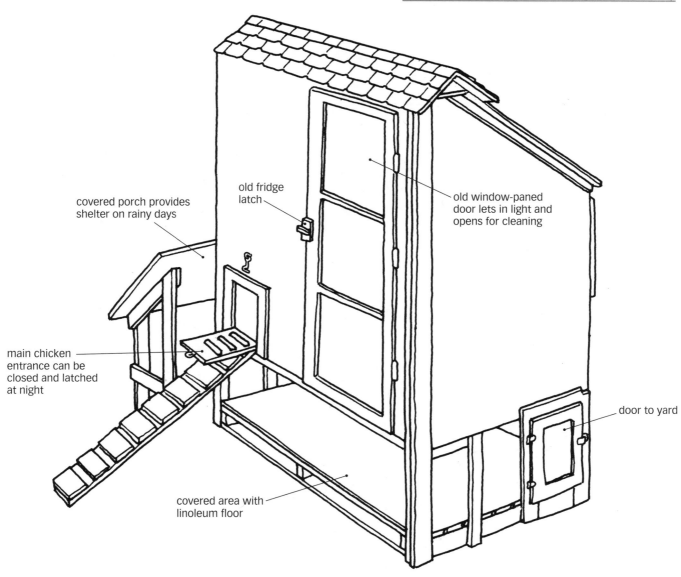

covered porch provides
shelter on rainy days

old fridge
latch

old window-paned
door lets in light and
opens for cleaning

main chicken
entrance can be
closed and latched
at night

door to yard

covered area with
linoleum floor

The Cat House (continued)

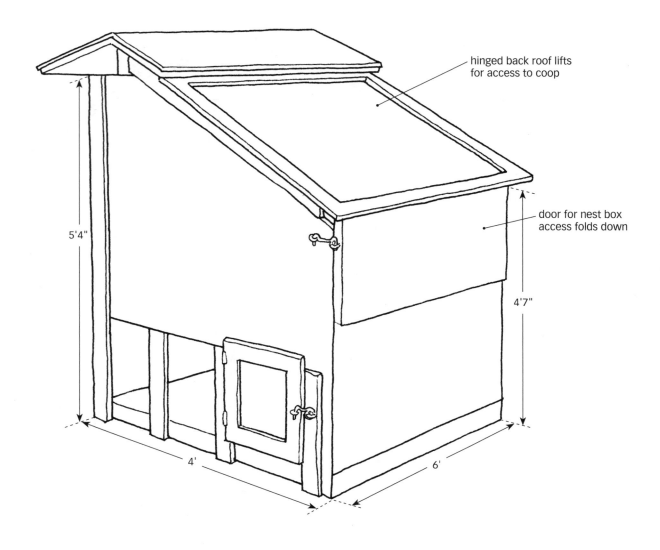

hinged back roof lifts
for access to coop

door for nest box
access folds down

5'4"

4'7"

4'

6'

Shake Your Tail Feathers Coop

(see color photo on page 153)

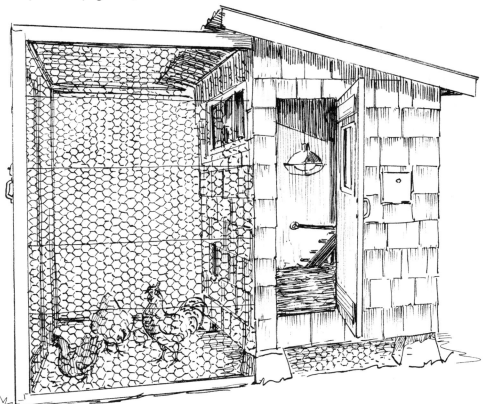

"I GREW UP IN Michigan and Montana and had chickens as a kid," says Jodi Clagg. "I moved to Seattle when I was 18, and I love living in the city, but I love the country, too." For Jodi and her husband, Mark, having chickens in their Seattle backyard is the best of both worlds. Mark and Jodi never imagined they could keep chickens in the city until she discovered a Seattle Tilth City Chickens class. After a few years of consideration, they settled on a coop design. Mark built the coop, and they purchased four chicks: an Araucana, a Golden Laced Wyandotte, a Buff Brahma, and a Partridge Cochin. Now Jodi is affectionately known in the neighborhood as the "Chicken Lady"! She is amused by the local interest in her four hens. "Neighbors bring their kids over to visit the chickens all the time," Jodi laughs. "People are constantly traipsing through the backyard to see what the girls are doing."

The coop is a freestanding 4-foot × 4-foot shed sided with cedar shakes. The door to the pen is a salvaged galvanized screen door covered with chicken wire. Both the door and the coop window were salvaged. The hens stay comfortable in the coop during the winter, and a heat lamp is needed only on the most extreme winter nights. The coop has one roost and a ladder to the nest box. The box can be accessed from a small door on the outside of the coop. A feeder is suspended from the coop ceiling. A waterer is located outside the coop. Raccoons are a problem, so the coop and pen are enclosed with buried chicken wire. There is an additional fenced area so the hens can day-range in the yard, which is accessed by a small door and ramp.

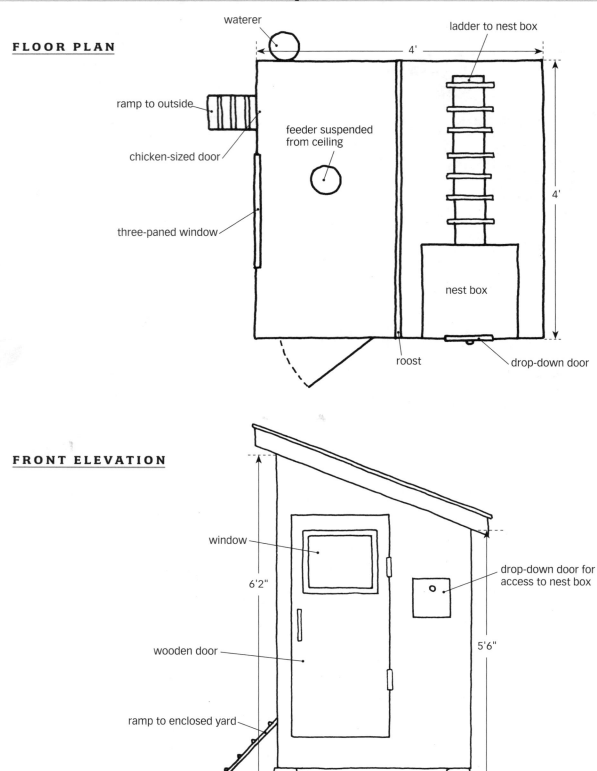

FLOOR PLAN

waterer

ladder to nest box

4'

ramp to outside

chicken-sized door

feeder suspended from ceiling

three-paned window

4'

nest box

roost

drop-down door

FRONT ELEVATION

window

6'2"

drop-down door for access to nest box

5'6"

wooden door

ramp to enclosed yard

Toolshed Henhouse

(see color photo on page 154)

HOLDS:
15 hens
3 nest boxes
3 roosts

CHICKENS CAME TO this Seattle backyard when Sharon Ely surprised her husband, Ray, with five chicks on his birthday. Ray grew up on a farm and later had his own chicken farm where he raised 25,000 chickens. Sharon figured city chickens would bring a little bit of the country to their backyard. After building two coops that allowed the birds to range free in the yard, the Elys began designing a larger coop that would keep the birds inside while allowing them to be seen. "My kids were old enough to be romping in the yard, and all the chicken doo that was being tracked around was annoying," Sharon explains. "I didn't want free-run chickens anymore."

The Elys progressed through a few different coop styles before deciding that they wanted the hens to have access to part of the yard. They also wanted the coop to be seen from a number of places in the yard as well as from the house. Ray and Sharon were designing a new coop when their toolshed was crushed in a snowstorm, so they created this 8-foot × 20-foot combination toolshed and henhouse. "I wanted the coop to be easy to maintain; otherwise I knew I just would not do it," says Sharon. "And we wanted to be able to see the hens."

The chicken coop has two sections, with an outside wall panel that can be picked up and set to the side for easy cleaning. "The main part of the coop is visible from our patio and hot tub and enclosed with light-gauge garden fencing," explains Sharon. This section holds the feeder, the waterer, two roosts, and three nesting boxes. The 3-foot × 6-foot indoor part of the coop has a roost and is insulated with rigid insulation board to keep the chickens comfortable during the mostly mild Seattle winters. A hose from the front of the building is run over the door and down into the automatic watering device on the front of the coop. The device works on pressure or gravity systems and can be found at some hatcheries (see page 160 for contact information). A flip top on the nest box makes it easy to remove the eggs.

"It was disturbing the first summer to see all the hens sprawled and motionless throughout the garden," says Sharon. "After a couple of false alarms of thinking they all had died, I got used to the idea that they actually were enjoying themselves." After the new coop was in use for a while, Sharon realized that a remodel was necessary. "The chickens were being deprived of their all-important dust baths. We built a small fenced-in pathway from the coop to an outside pen where they can scratch, bathe, and eat worms."

FLOOR PLAN

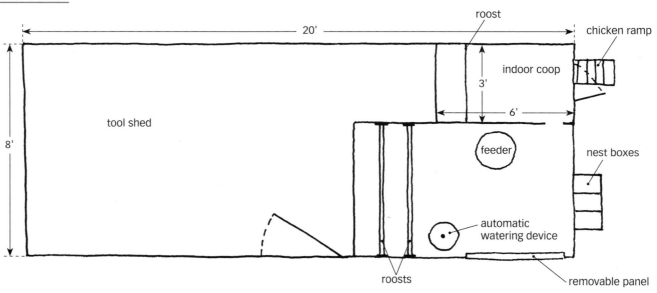

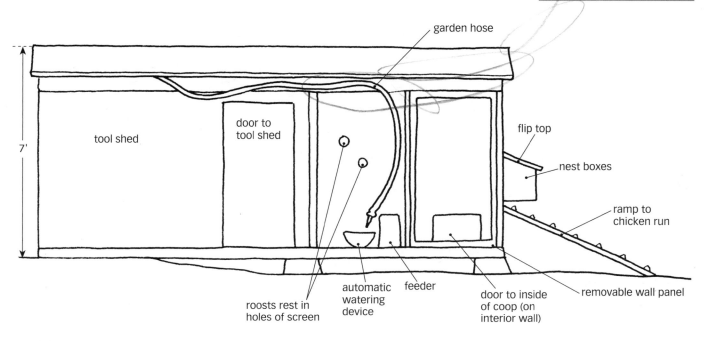

garden hose

7'

tool shed

door to
tool shed

flip top

nest boxes

ramp to
chicken run

roosts rest in
holes of screen

automatic
watering
device

feeder

door to inside
of coop (on
interior wall)

removable wall panel

VACUUM-STYLE PLASTIC WATERER AND OLD-FASHIONED METAL FEEDER

These very common feed and water containers are from Shiloh Farm and Retreat. They both can be found new in most farm-supply stores. This feeder, with ample room for food for several chickens and a handle that can be hung, has been a standard for years. Old feeders can still be scrounged from farms that used to produce chickens.

waterer

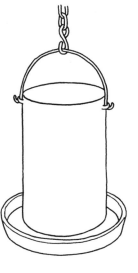

feeder

Toolshed Henhouse (continued)

SIDE ELEVATION

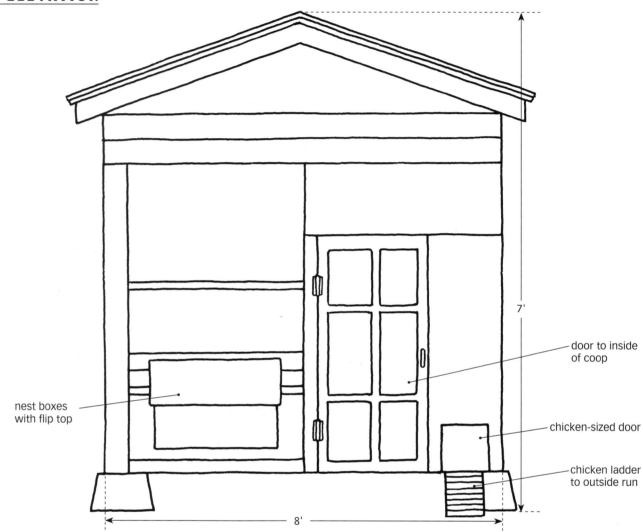

nest boxes
with flip top

door to inside
of coop

chicken-sized door

chicken ladder
to outside run

7'

8'

Starclucks Coop

(see color photo on page 154)

CRANE STAVIG was looking for a new project to work on last year. Recently laid off, he had plenty of time but not a lot of money. Then Crane attended the Northwest Flower and Garden Show in Seattle and discovered Seattle Tilth's City Chickens classes. The rest is Starclucks history!

"I have a large yard and a young daughter, Lena, and I enjoy gardening, so it sounded like a fun way to add some low-maintenance pets to the landscape that would also give us fresh eggs and manure," Crane says. "It sounded like a winning proposition all the way around." After attending the City Chickens 101 class, researching chickens on the Internet and in books, talking to (warning) neighbors, and considering housing requirements, Crane started sketching plans. The design started as a simple chicken coop, but

then he decided he needed a garden shed. When Crane's salvaging was more successful than he had anticipated, he added more windows and enlarged the coop.

As the coop design was evolving, Crane bought seven day-old chicks. After all, he reasoned, how fast could they grow? He had his answer soon as they quickly outgrew the cardboard box Lena had decorated. Crane rushed to get the coop built. After eight weeks, the Starclucks Coop was finished, and the chicks had their permanent home. During the City Chickens Tour, Crane strongly recommends that anyone planning on raising chickens makes sure that he or she has built, or at least has started, the coop before bringing the chicks home!

The Starclucks Coop is designed to hold as many as seven heavy-breed laying hens.

Insulated with fiberglass insulation, the henhouse easily withstands Seattle's mild winters. Half of the components used to build the coop are recycled and salvaged materials. A rain barrel on a 3-foot-tall wooden stand behind the garden shed gathers water to feed the automatic waterer near the front of the coop. A rain gutter that runs the length of the roof drains down a spout and into the rain barrel. The water then runs via a garden hose along the interior wall of the garden shed to the automatic waterer.

The coop has a total of 16 feet of roosts inside the henhouse and in the covered run. To make for easy removal and cleaning, each roost is mounted with fence-rail brackets and no fasteners. The hens share two nest boxes that are cantilevered off the back of the henhouse to allow more space inside. A lid on the nest box allows for easy access, and a clasp on the lid allows it to be locked at night to keep critters out. The coop and shed have a translucent roof that lets in additional natural light.

During the summer, when the chickens are not peeking at Crane as he putters in the garden shed, they scratch in the soil and dust themselves in the attached outdoor run. The run is enclosed by chicken wire to protect the chickens from predators. Roosts in the outside run are a unique feature and receive a lot of attention during the City Chickens Tour. They are the hens' favorite spots to nap and soak up the sun.

FLOOR PLAN

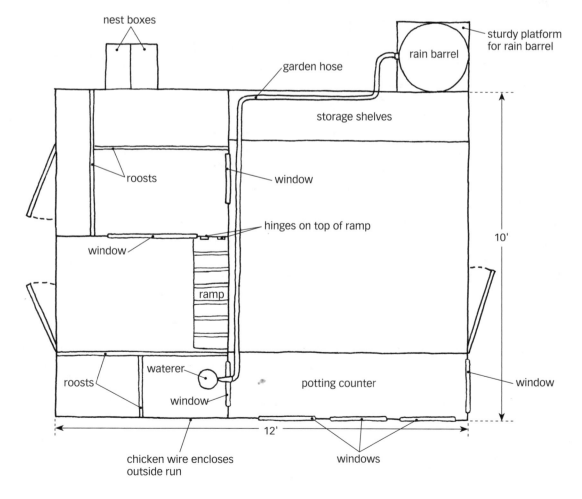

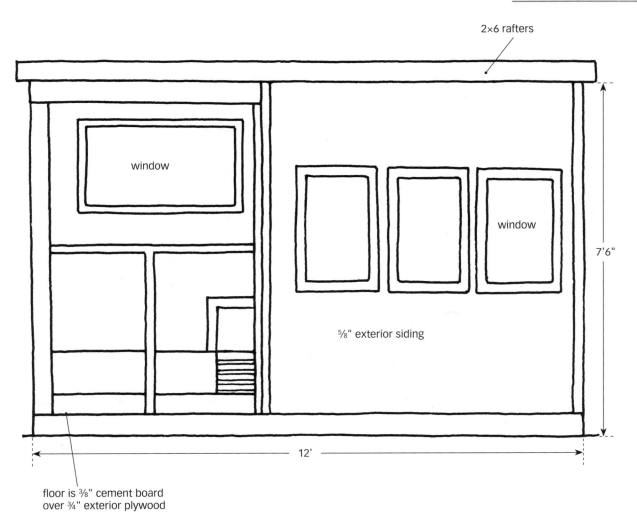

2×6 rafters

window

window

⅝" exterior siding

7'6"

12'

floor is ⅜" cement board
over ¾" exterior plywood

Starclucks Coop (continued)

SIDE ELEVATION

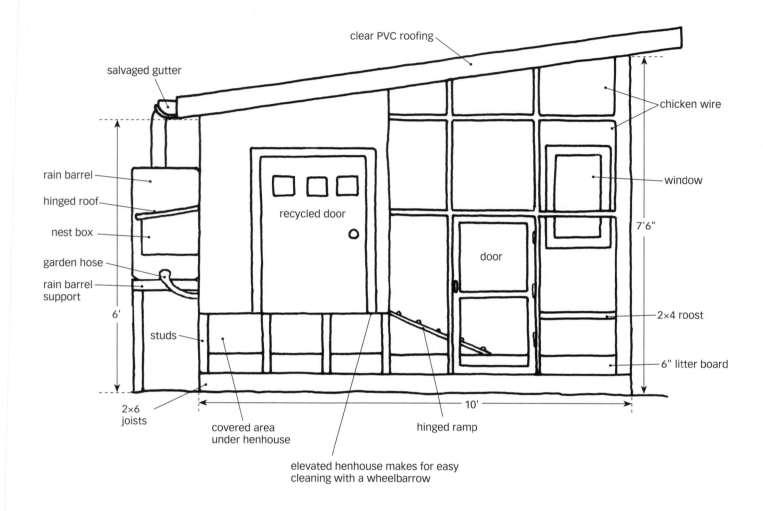

clear PVC roofing

salvaged gutter

chicken wire

rain barrel

hinged roof

window

nest box

recycled door

7'6"

garden hose

door

rain barrel support

2×4 roost

studs

6

6" litter board

2×6 joists

10'

covered area under henhouse

hinged ramp

elevated henhouse makes for easy cleaning with a wheelbarrow

Kids' Garden Coop

(see color photo on page 153)

CHILDREN FROM ALL OVER Madison, Wisconsin, hang out with the hens at the Kids' Garden Coop and fall in love with raising chickens. Tobias Harrison-Noonan built this coop to complete his Eagle Scout community-service project. He modeled it after a coop his father had built to house his four heavy-breed layers. Toby and his Boy Scout troop gathered donated and salvaged materials. When the coop was finished, they installed it in the middle of the children's section of the local community garden.

The coop is made with cedar and plywood siding and has a 4-foot × roughly 8-foot run. The roof is a trendy agricultural-style metal that blends well with neighboring houses. Four people can lift the coop and move it around the garden as needed.

The nest box is a 12-inch × 12-inch wooden cube open on two sides. One side allows the hens to enter from inside the roost, and the other opens to the padlocked egg-collection door on the outside. The roost bar passes over the top of the nest box, so the nest-box roof comes in handy by shielding the box from the roosting birds' droppings. A hinged door that opens from the top down on the interior side of the nest box/roost area provides good ventilation in the summer, when it can be folded down. When it is closed in winter, the door also provides extra warmth. A sliding plywood door on the side of the roost allows access for bedding changes and for cleaning the nest box/roost area. Plans with complete specs can be purchased from Dennis Harrison-Noonan (see page 162 for Web address).

Kids' Garden Coop (continued)

FLOOR PLAN

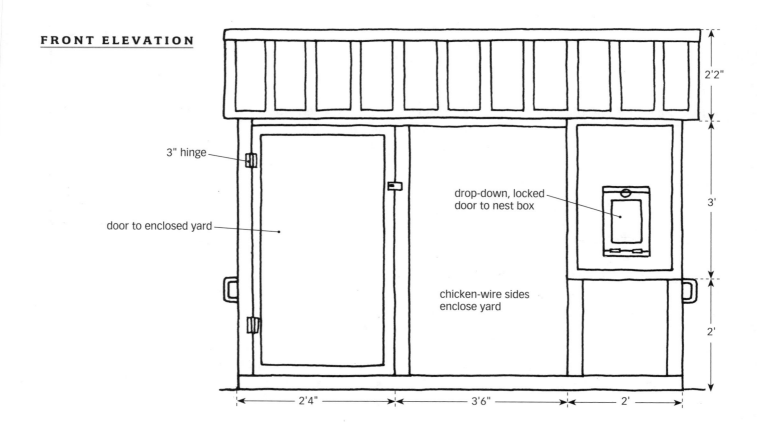

7'10"

chicken ladder

sliding door

roost area

4'

door

window

feeder

waterer

door swings down

nest box area

door for egg collection

2'4" door

FRONT ELEVATION

2'2"

3" hinge

drop-down, locked door to nest box

3'

door to enclosed yard

chicken-wire sides enclose yard

2'

2'4"

3'6"

2'

CONSTRUCTION NOTES

▶ The 4-foot × 8-foot bottom is made of galvanized welded-wire mesh fencing stapled to the bottom of the cedar frame.

▶ The frame is made of ⁵⁄₄" cedar deck boards ripped to 2½".

▶ The roost and nest box area has a sliding plywood door to make cleaning easier.

▶ The plywood door across the nest box/roost area is hinged with two 5-inch hinges.

▶ The sides are made from 1" × 1" galvanized wire mesh.

▶ Siding is made of ⅜" rough fir plywood.

SIDE ELEVATION FACING OUTSIDE

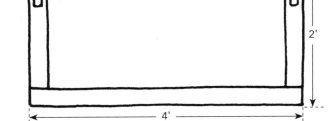

prefinished galvanized steel roof

16"

window

flower box

metal handle

3'

2'

4'

SIDE ELEVATION FACING INSIDE

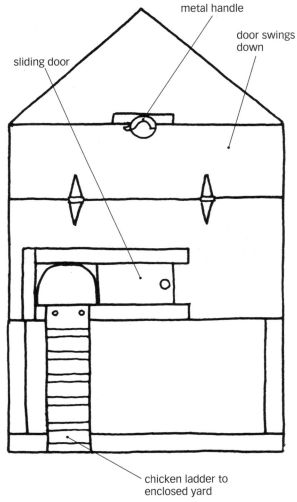

metal handle

door swings down

sliding door

chicken ladder to enclosed yard

Chicks 'n the 'Hood Coop

(see color photos on page 152)

HOLDS:
4 hens
4 nest boxes
2 roosts

THE ENTIRE NEIGHBORHOOD has adopted the Bassler-Mortensen family's hens in Madison, Wisconsin. "The hens spend most of their time free-ranging in our yard and garden," says Karen Bassler. "They foray out to neighbors' yards quite a bit and thus have become more or less adopted by the block." Neighbors wander over to the backyard to talk to the girls or just to watch them. Little Alia, age one and a half, calls them "Cheeek." Next-door neighbor Ken is the hens' unofficial guardian; he informs the family of any disturbances or stalkings that occur during the day. "They wander among the various backyards, taking dust baths in Pauline's garden beds, kicking out all the wood chips from around Pat's new redbud tree, and frolicking in the compost pile behind their coop. Shara and Egon [my children] keep watchful eyes on them and herd them back into the coop when it is time. Morticia is the only hen that is amenable to being caught. She just sits and lets you pick her up. Jemima is jumpy and skittish and impossible to catch. Shirley will tolerate being held, and Peep on occasion lets the kids pick her up."

Egon makes "chicken cake" for the hens: a mixture of yogurt, smashed garlic, oats, cornmeal, matzo meal, and chicken feed garnished with greens plucked from the garden. Both Egon and Shara gather the eggs. When it is time for the family's Monday-night dinner exchange with two other families, there are six children "running amok in the backyard," Karen laughs. "The birds love the chaos, and the kids love the birds and their antics."

The coop is made from recycled and salvaged materials. It is secured by a wire pen with a door that shuts closed at night. The hens snug into four nest boxes on the side of the coop and roost on two pegs that stretch across the top. The hens can jump down from the roosts and nest boxes and hang out in the bedded area below or wander outside through the hen door on the side. The other side has a door that opens to the side so that the eggs can be collected from the outside. One side of the roof is hinged so that it can be lifted to allow access to the entire coop for cleaning.

MAD CITY CHICKENS TOUR

PEOPLE ARE MAD about chickens in Madison, Wisconsin, now that a city ordinance allows residents to keep up to four hens in their backyards. Mad City Chickens is a group dedicated to sustainable agriculture and healthy eating. The aim of its members is to educate city dwellers on the benefits of raising chickens. They offer an introductory class for those urban- and suburbanites interested in raising birds in their backyard. They enjoy the support of the University of Wisconsin's Poultry Department and are eager to answer questions. Their Web site (see page 162 for Web address) provides chicken-related classifieds, answers to frequently asked chicken questions, and links to chicken-related Web sites.

These city chickens are housed in small, smart-looking coops to eliminate resistance from those few residents who view the birds as smelly, noisy farm animals that will have a negative impact on property values. Attractive coops and keeping the noise down by not having roosters (sorry, boys) keeps the peace with the neighbors.

FLOOR PLAN

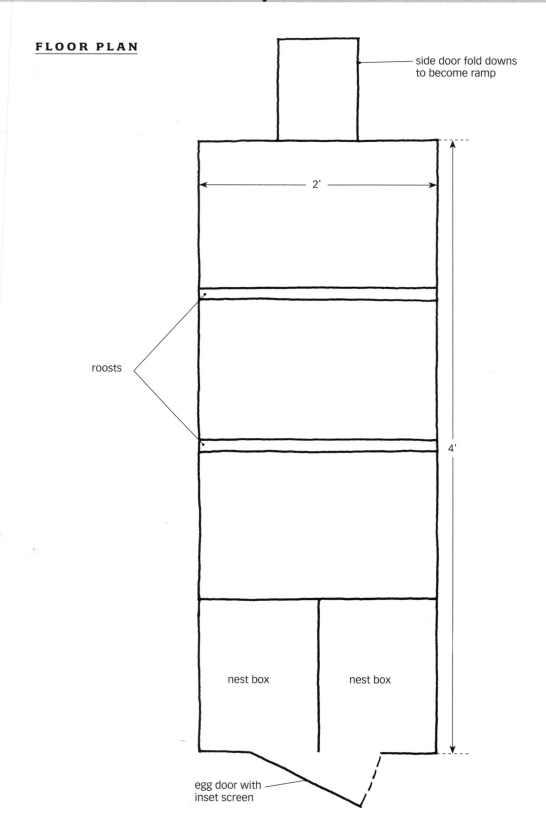

side door fold downs
to become ramp

2'

roosts

4'

nest box   nest box

egg door with
inset screen

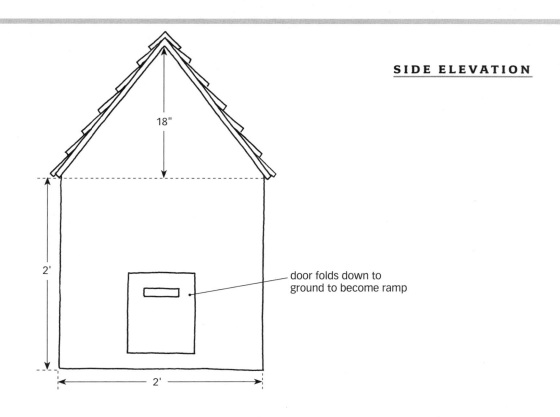

18"

2'

2'

door folds down to
ground to become ramp

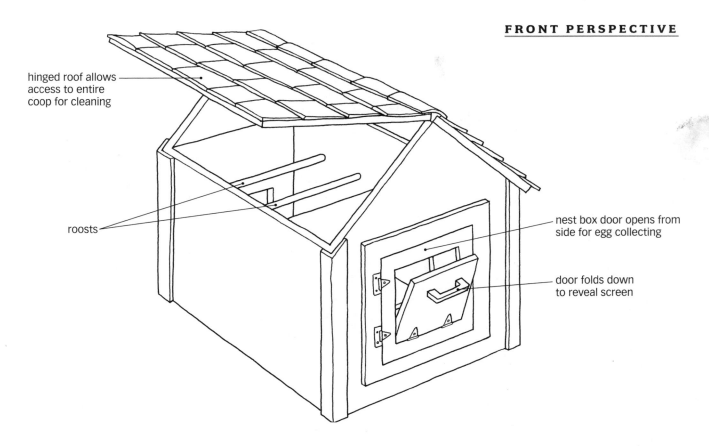

hinged roof allows
access to entire
coop for cleaning

roosts

nest box door opens from
side for egg collecting

door folds down
to reveal screen

3 Coops for Small Farms

IN THIS SECTION, we've included not only coops for laying hens but also coops for broilers and turkeys, due to the rising popularity of pastured poultry. Consumers turned off by factory farms are seeking safe, natural, and healthy chickens and eggs from local family farms. Animal-rights advocates insist that chicken-production models should give the birds plenty of space, fresh air, and feed free of chicken parts and antibiotics. To answer these demands, farmers are raising broilers and laying hens in mobile chicken coops that are moved often so the birds have free access to fresh green pastures.

The consumer interest in pastured poultry has sparked a new interest in farming in general. The average farmer in the United States is 55 years old, and land prices have increased dramatically in areas where development pressure is high. Raising pastured poultry is attractive to farmers of all ages because it quickly generates revenues while requiring only limited acreage and demanding little in the way of machinery and labor. Poultry housing can be built from salvaged and recycled materials. Many new, but not necessarily young, farmers can work off-farm jobs and still raise pastured poultry as they develop markets for broilers and eggs. Experienced and new farmers alike are discovering that a pastured-poultry operation can be a stand-alone enterprise on a new farm, an added enterprise, or the first step in converting the farm from a conventional dairy, crop, or livestock farm to a sustainable grass-based farm. When properly marketed, meat birds and eggs will bring customers to the farm from many miles away.

The coops in this section were built for small-scale chicken or egg production. They are meant to be inexpensive and are designed to allow very low labor input in the care of the chickens. Many of the coops were built from materials at hand on the farm, and a few are recycled farm buildings or wagons. Most of the coops allow the chicken manure to drop straight to the ground, reducing the amount of labor required to clean the pens. Nesting boxes are easily accessible to allow for fast and easy egg collection. Long-term feed and water storage is either in the coops or located nearby. If you are interested in starting a small-scale hen or chicken production operation or are expanding your existing operation, the coops and equipment in this section will provide inspiration.

Truck Cap Coop

SEEMS LIKE YOU CAN always find a truck cap lying around, free — or almost free — for the taking! Set it up on some blocks or plywood sidewalls, hang a nest box inside one end, add a waterer and some feeders, surround it with portable netting, and you have a ready-made summertime chicken coop. This coop does not have roosts, but they could easily be installed across the inside of the truck cap and attached to the wood sidewalls. The hens use a 15-unit nest box that is moved over to the Egg-loo (see page 76) in the winter.

Eric and Melissa Shelley and Eric's sister, Dr. Cindi Shelley, own this Truck Cap Coop. Eric and Melissa own and operate the only New York State–regulated mobile meat-processing unit in the area surrounding Sharon Springs, New York, a town about 45 miles west of Albany. Cindi is a professor of animal science at the State University of New York College of Agriculture and Technology at Cobleskill. In their spare time, Eric, Melissa, and Cindi are renovating a farmhouse and raising dairy goats, laying hens, and grass-fed beef.

FLOOR PLAN

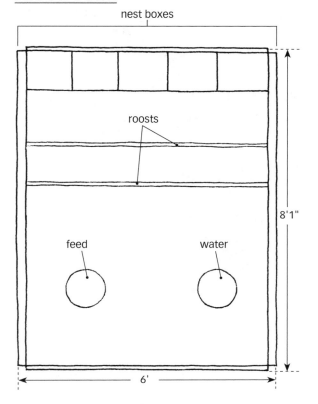

nest boxes

roosts

feed

water

8'1"

6'

FRONT ELEVATION

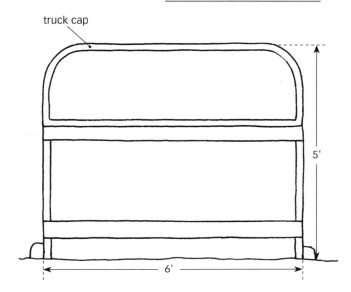

truck cap

5'

6'

THE SHELLEYS' MOBILE MEAT-PROCESSING UNIT

Eric Shelley's mobile meat-processing unit serves small local livestock farms, including ours. To use Eric's unit, we had to obtain a food-processing license from the state. After our beef and pigs have been slaughtered at an off-farm facility inspected by the USDA (United States Department of Agriculture), Eric rolls in the truck and plugs into the farm's electric and water systems, and he and Frank cut and wrap all of our meat.

Using the mobile processing unit gives us complete control of every cut and allows us to make our own specialty sausages and other value-added products. Under this system, we are allowed to sell the meat directly to customers at the farm and at farmers' markets, and also to restaurants, all within New York State. Funding assistance for the processing facility in the form of a low-interest loan was provided by the Schoharie County Industrial Development Agency.

Egg-loo

WHEN WINTER COMES to Eric and Melissa Shelley's farm in Sharon Springs, New York, the Truck Cap Coop is closed and the hens are moved next door to the Egg-loo. Old rectangular hay bales are stacked in a circle much like ice blocks stacked to form an igloo.

The long sides of the coop consist of three hay bundles placed tightly together, with the third bundle angled toward the inside of the coop. Two bales are laid across lengthwise to form one end. A gap is left between the ends of the angled sides to create a hen door. Half bales are inserted at the ends where needed to make the walls airtight. Three sturdy tree limbs are placed across the center of the hay bundles on the second level and double as roosts for the hens. Another set of braces is placed over the third layer of bundles to hold the roof bales. Two vertical tree limbs support the horizontal limbs in the center. The roof bales are covered with an old tarp.

A 15-unit metal nest box is tucked between the hay bundles on one end. Originally, the outside of the nest box was covered with a sheet of rigid-board insulation that could easily be moved aside for egg collection from the outside. Unfortunately, the hens liked to chew on the insulation, so it was removed and hay bales were used instead. These are moved out of the way for egg collection. The feeder is hung on the inside from one of the limb braces. A small gap is left in the bundles of hay just above the feeder so that Eric can stand on the ground outside of the coop and pour the feed from a bucket between the bales and into the feeder. Eric used old hay bundles that were baled slightly wet. These will heat up over the winter and provide extra warmth in the coop. A bale is broken and shaken across the floor of the coop for extra bedding. Electric netting surrounds the coop and provides predator protection.

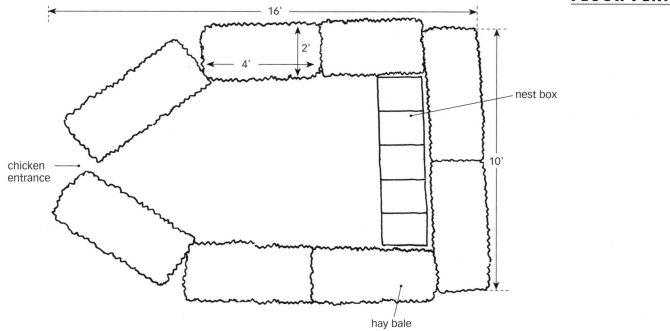

16'

2'

4'

nest box

10'

chicken
entrance

hay bale

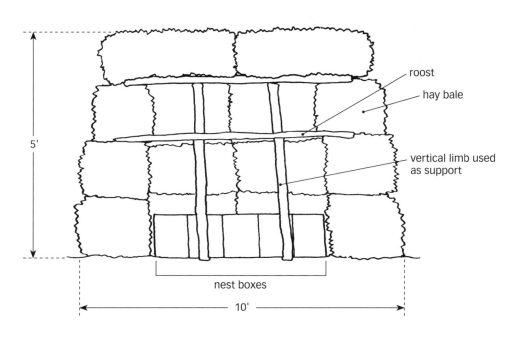

roost

hay bale

5'

vertical limb used
as support

nest boxes

10'

Beck-Chenoweth Broiler Skid

HOLDS:
300 broilers

AT HIS FARM IN OHIO, Herman Beck-Chenoweth developed a free-range poultry-production system for broilers. In it, a series of 8-foot × 18-foot skid houses that hold up to three hundred birds each are placed 100 feet apart in the pastures. They are towed by tractor every few weeks to new pasture.

The floors are wooden skids enclosed with chicken wire and covered with litter. The gable roofs can be covered with tarps or plastic or metal roofing material. The front and back walls of the house lay flat on the ground during the day and are tied upright against the walls at night. (The walls can also be built with a regular door instead of the drop-down feature.) A secure perimeter fence or electric poultry netting can surround the entire setup. The meat birds free-range about 40 feet away from the skids during the day and are closed in at night for added predator protection. Keeping the skid houses at least 100 feet apart ensures that the birds won't confuse one house for another, thus making sure that they return to their own house at night. Water is provided by automatic watering systems in the pastures.

"The low stocking rate, four hundred chickens or one hundred turkeys per acre, allows the birds plenty of area to exercise and deposit manure," says Herm. "As a result, free-range birds develop excellent muscle tone. Since the muscle is what we eat, this development is very important. Combined with proper aging after slaughter, meat quality is firm but smooth — second to none." This method is described in detail in his book; Herm also provides chicken information on his Web site (see page 162).

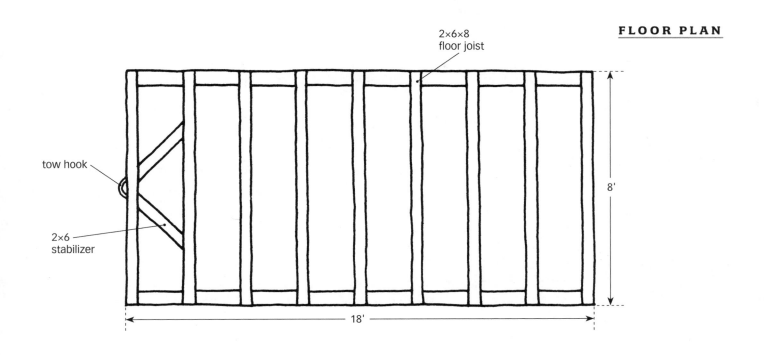

2×6×8
floor joist

tow hook

2×6
stabilizer

8'

18'

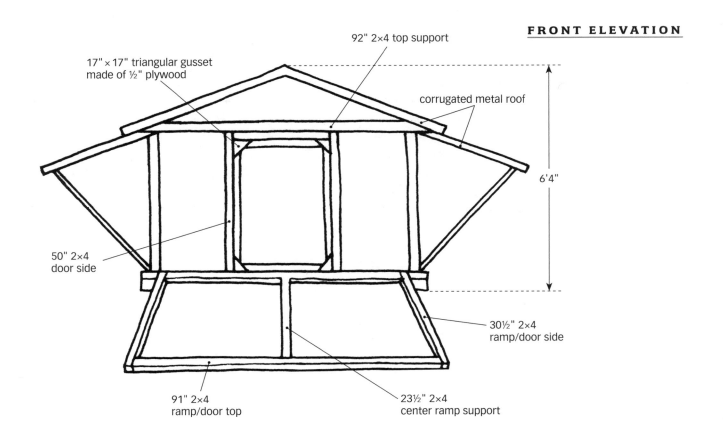

92" 2×4 top support

17" × 17" triangular gusset
made of ½" plywood

corrugated metal roof

50" 2×4
door side

6'4"

30½" 2×4
ramp/door side

91" 2×4
ramp/door top

23½" 2×4
center ramp support

CARETAKER FARM

JUST OUTSIDE WILLIAMSTOWN, MASSACHUSETTS, Elizabeth and Sam Smith started Caretaker Farm, one of the earliest organic farms in the Northeast. The farm produces a variety of organic vegetables, fruit, and flowers, as well as beef, pork, lamb, and eggs. Elizabeth was always the "animal person," caring for the cows, pigs, sheep, and chickens, and Sam grew six acres of vegetables. In 2006, the future of Caretaker Farm was secured in a community land trust to keep it as a working farm affordable to future farmers in perpetuity. While the responsibility for the farm has passed into the hands of younger farmers, Sam and Elizabeth continue to live on Caretaker Farm as the farmers emeritus.

The Smiths are leaders in the sustainable-farming movement, with a primary focus on education and building community through local food production. Caretaker Farm is a community-supported agriculture (CSA) model in which members of the CSA prepurchase shares of the following growing season's production. This allows CSA members to share in the harvest and in the risks inherent in farming and provides the Smiths with operating capital for the coming growing season. During the growing season, members share Caretaker Farm's weekly harvest. To build community among the operation's CSA members, the Smiths encourage members to work at Caretaker Farm, help with harvesting, and become part of the seasonal rhythms of life on the farm. Through these efforts, the farmers and CSA members alike are committed to the enterprise's success.

The Smiths believe that soil fertility is the key to Caretaker Farm's sustainability. To improve the farm's soil fertility, the vegetable fields are rotated with a cash crop one year and cover crops the next. Laying hens were added to further improve soil fertility. The laying hens are free-ranged over fallow fields for the nutrients in their droppings and to control slugs and other bugs. They work the compost piles by scratching for bugs in the top layers and encouraging decomposition. The mobile chicken coop is moved behind the beef in the fields, and as they scratch through the cow manure searching for bugs, the chickens work the manure directly into the soil for instant organic matter. The laying-hen enterprise is a win-win for both the farm and the CSA members, who are thrilled with the quality of the fresh eggs they receive.

Chicken Prairie Schooner

(see color photo on page 155)

HOLDS:
50 hens
1 nest box
6 roosts

WHEN SAM AND ELIZABETH SMITH of Caretaker Farm decided to add pastured laying hens to the crop rotation on their certified-organic vegetable farm, they knew that the coop had to meet certain parameters to be functional. Because the farm is run without machinery, the coop had to be light and mobile enough to be easily moved by hand over hilly terrain. It had to withstand strong winds. It had to fit through the garden gates. The hens and all their accoutrements had to fit inside the coop whenever it was moved.

After trying a few styles of mobile coops, Elizabeth built her first prairie schooner. The schooner proved to be a success, though it was somewhat heavier than she would have liked. After using the schooner for a time, Elizabeth built a slightly narrower version (plans shown are for this coop) that was lighter and easier to move. Both schooners are now used side by side to house the farm's flock of laying hens.

The Chicken Prairie Schooner features metal wheels salvaged from old horse-drawn hay rakes. Because they have become popular as lawn decorations and are now difficult to obtain, Elizabeth is always on the lookout for these wheels. The hoops are plastic PVC pipes anchored on steel rebar. The schooner's bonnet is heavy plastic salvaged from a lumberyard.

The plastic-coated chicken-wire floor is easy on the hens' feet and lets the chicken droppings fall to the ground. The coop features a community nesting box that is accessed by people from the outside. The top of the box flips up so that eggs can be collected from outside the schooner. Automatic waterers are gravity-fed from pails hung from the back of the schooners. The ramp on the back provides access to the coop for the hens and flips up to close them inside when it is time to move. Handles across the front of the coop are used to move the schooner to greener pastures.

Chicken Prairie Schooner (continued)

CONSTRUCTION NOTES

▶ Coop can be made smaller (4' × 8') so that it is easier to manuever.

▶ Construct 2×4 frame with deck screws 18" apart.

▶ Cover PVC hoops with plastic-coated chicken wire. Heavy plastic bonnet goes directly over chicken wire.

▶ There are a total of six roosts, each 2" × 2" × 5'3½". Each bird has 8 inches of roost space.

▶ Drill holes into top of 2×4 frame to peg in ¼" rebar spikes 3 feet apart. Thread PVC pipes onto spikes.

▶ Secure door frame to hoop with wire straps.

▶ Roosts are integrated into the frame, on top of the wire, and placed at roughly 18" intervals.

FLOOR PLAN

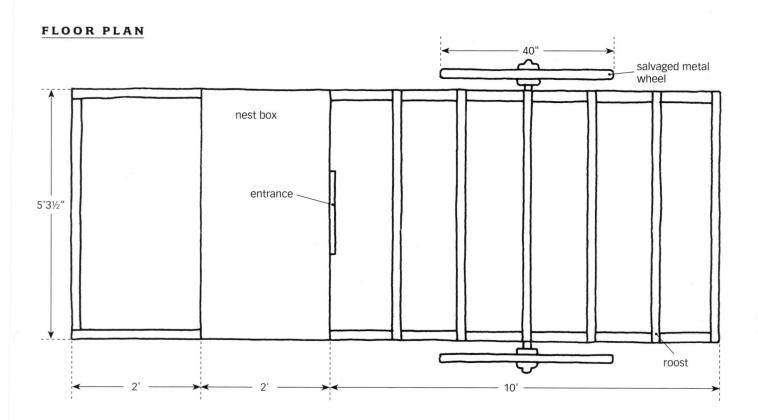

salvaged metal wheel

nest box

entrance

roost

40"

5'3½"

2'   2'   10'

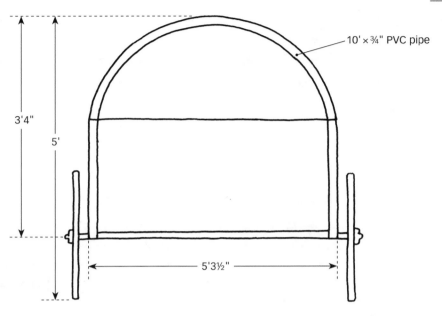

3'4"

5'

10' × ¾" PVC pipe

5'3½"

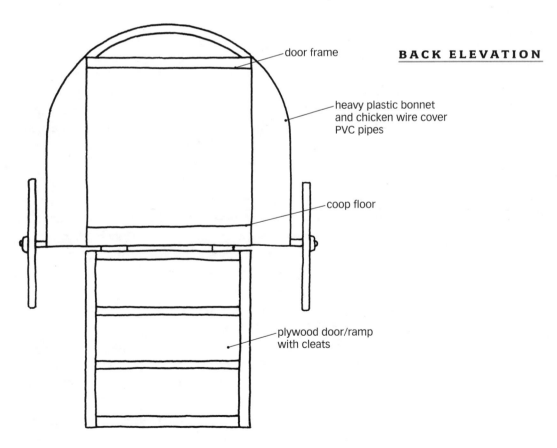

door frame

heavy plastic bonnet
and chicken wire cover
PVC pipes

coop floor

plywood door/ramp
with cleats

SAP BUSH HOLLOW FARM

WE VISITED SAP BUSH HOLLOW FARM in West Fulton, New York, on a beautiful day when the farm was awash in brilliant Catskill Mountains fall colors. We found Jim and Adele Hayes processing a batch of broilers at the farm's processing facility. While we waited for them to finish, our son Arleigh tossed a disc to one of the farm's nine sheep and guard dogs, and Frank and I took a break and simply enjoyed the view.

The scene there was an old-fashioned farm painting come to life. Laying hens scratched in the grass, and pigs rooted by the barn. Brilliant green fields stretched across the valley, with turkeys in the foreground and sheep, broilers, and Jim and Adele's two guard dogs in the background at the base of the mountains. Beef steers peeked through the trees behind the turkey pen. Guinea hens pecked in the grass next to the driveway, and a small flock of geese waddled across the yard and peered in at Jim and Adele as they worked. The final touch to this pastoral scene was the black kitten that gamboled at our feet as we walked the pastures.

Jim and Adele Hayes, along with their daughter and son-in-law, Shannon Hayes and Bob Hooper, raise grass-fed sheep, beef, pork, broilers, turkeys, and laying hens. "Our animals enjoy fresh air, sunshine, lush green grass, and fresh water; they live a stress-free life," explains Adele. They sell all of their products right from the farmhouse. Many customers make the trek to Sap Bush Hollow Farm every few weeks for sale days; some come from as far away as Connecticut, Massachusetts, and New York City. Chickens are available by preorder at sale days held every four weeks during the summer; frozen meat cuts are also available. Meat sales continue through the winter and are held at least once a month. Turkeys are available by order for Thanksgiving. Customers must come to the farm to pick up their preordered fresh chickens and turkeys. In addition, Jim and Adele are forming a buying club in the Albany/Schenectady area, located about 50 miles away. Bob and Shannon and their daughter also sell the farm's products at the Pakatakin Farmers' Market in Margaretville every Saturday in the summer.

Hayes family members mentor beginning grass farmers and are instrumental in educating consumers about the benefits of pastured poultry. They speak at grass-farming and sustainable-agriculture workshops and seminars across the nation. Shannon earned her Ph.D. in sustainable agriculture and community development from Cornell University and has also published a cookbook on cooking grass-fed meat entitled *The Grassfed Gourmet* (Ten Speed Press, 2005). In addition, the family holds direct-marketing classes on the farm, where participants spend two days immersed in learning how to produce and market grass-fed livestock and poultry.

When our family began farming, we were the definition of "green" grass farmers. Jim and Adele have been our mentors since we began. We owe a great deal to the Hayes family for sharing their knowledge and friendship with us over the years.

Sap Bush Broiler

(see color photo on page 155)

& Turkey Hoop

HOLDS:
225 to 450
broilers in
four hoops

CREATING A MEAT-BIRD HOUSE or turkey house can be as simple as bending a cattle panel into the shape of a hoop, connecting it to a wooden frame, and covering it with a tarp. After trying a number of coop styles for their broilers, Jim and Adele Hayes chose these small hoop houses for their turkeys and broilers for a number of reasons: the hoop house is particularly apt for hot summer days when the birds quickly become stressed in other, less-ventilated coop styles; the coop is easy to move without harming the birds; and the birds can roam in and around the coops while still being protected from predators by the surrounding poultry netting.

The coop is 8 feet × 10 feet and consists of two 16-foot-long cattle panels tied together with hog rings and attached to a wooden frame. The structure has no floor.

The ends of the coop are made from cattle panels cut to the width of the pen. They are secured with ties to the hoop frame at night and untied to lie flat on the ground during the day. The open area between the top of the hoop and the top of the end panels is covered with construction safety fence to keep owls at bay while allowing open ventilation. A small tarp can be dropped over each end for shade. Chicken wire covers the bottom 2 feet of the coop. A 10-foot × 16-foot tarp to cover the entire hoop can be rolled up or down depending on the weather.

Four coops, all surrounded by battery-powered poultry netting, easily hold between 225 and 450 broilers. The chicks are raised in the brooder for three weeks and then placed in the hoop houses. Turkey poults are raised in the brooder and then placed in separate hoop houses for two weeks before being moved to the turkey roost.

When it is time to harvest the broilers, Jim and Adele set chicken crates on their sides to form a barricade between two hoop houses. The birds are herded into this temporary pen, placed in chicken crates, and brought to the farm's open-air facility for processing. This method is a vast improvement over the old knee- and backbreaking method of trapping the birds in the coops and then loading them into the crates!

FLOOR PLAN

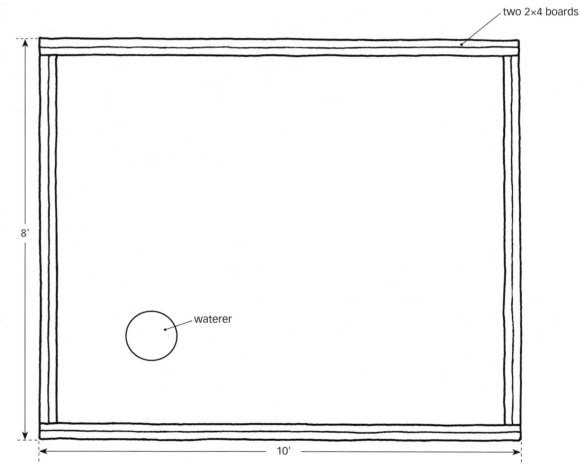

two 2×4 boards

8'

waterer

10'

CONSTRUCTION NOTES

▶ There are a number of ways to secure the panels to the wooden frame. The panels can be sandwiched between two boards that are screwed together, for example. Or each panel could be connected to a single board with fence staples, bent nails, or strap-style pipe fasteners (the kind that are placed over the wire and can be screwed down on either side).

▶ The 16' cattle panels are connected with hog rings.

▶ The cattle panels on the ends are attached to the house with ties. The panels on both ends lie flat on the ground during the day.

END ELEVATION

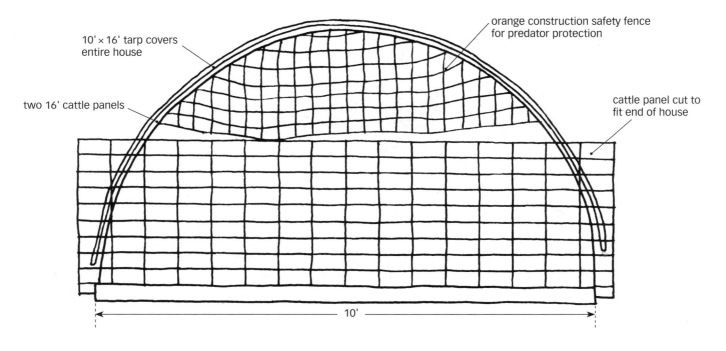

10' × 16' tarp covers entire house

two 16' cattle panels

orange construction safety fence for predator protection

cattle panel cut to fit end of house

10'

Sap Bush Turkey Roost

HOLDS:
75 to 100
turkeys
7 roosts

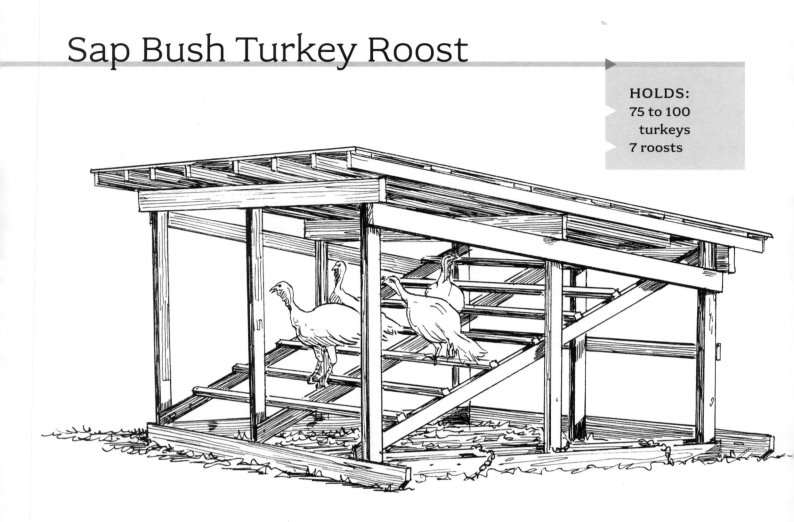

THE TURKEY ROOST at Sap Bush Hollow Farm in West Fulton, New York, is a wood-framed open shed with a metal roof. Built on skids, the structure is moved one coop length every day by hooking a chain from the tractor to short chains attached to the bottom skid. Battery-powered poultry netting surrounds the coop and provides predator protection. The entire turkey pasture is enclosed by a permanent block-wire sheep fence. The turkeys are moved from the hoop houses to the turkey roost when the birds are about five weeks old. After a few weeks in the turkey-roost pen, the netting is opened and the turkeys are given access to the entire pasture. To avoid cross contamination with the chickens, the chickens are never allowed on the turkey pastures.

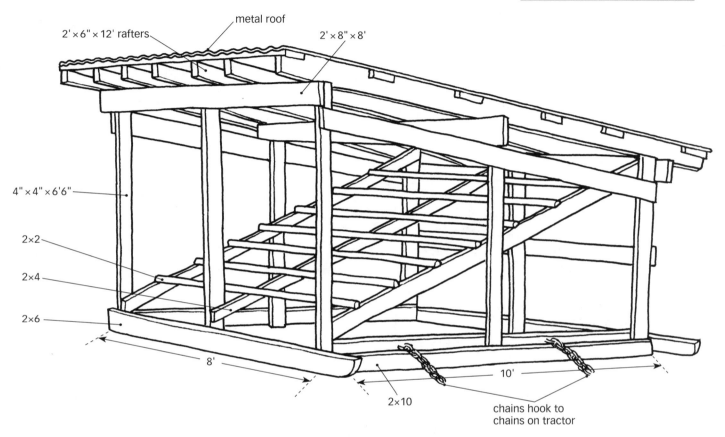

2' × 6" × 12' rafters

metal roof

2' × 8" × 8'

4" × 4" × 6'6"

2×2

2×4

2×6

8'

2×10

10'

chains hook to
chains on tractor

TURKEY ROOSTS VS. CHICKEN ROOSTS

Because turkeys are heavier and larger than chickens, turkey roosts need to be slightly farther apart and sturdier than chicken roosts.

Sap Bush Laying-Hen Coops

THE LAYING HENS on Sap Bush Hollow Farm in West Fulton, New York, spend their nights in a mobile Chicken RV and lay their eggs in a separate Eggmobile. "Keeping the sleeping quarters entirely separate from the nest boxes results in pristine eggs," Adele explains. "We open the Eggmobile in the morning and close the RV so the hens must go into the Eggmobile to lay their eggs." The coops are moved together across the pastures behind the sheep and beef and are surrounded by battery-powered poultry netting for predator protection.

HOLDS:
125 hens
20 nest boxes (Eggmobile)
2 roosts (Chicken RV)

Sap Bush Chicken RV

The Chicken RV is built on the frame of an old GMC truck. The sides and roof are sheets of translucent roofing material. Two roosts stretch across the width of the inside of the coop. The door on the back can be closed at night to protect the chickens from predators and to keep them enclosed during early-morning moves. The RV is cleared of hens and closed in the morning so the hens cannot access it during the day.

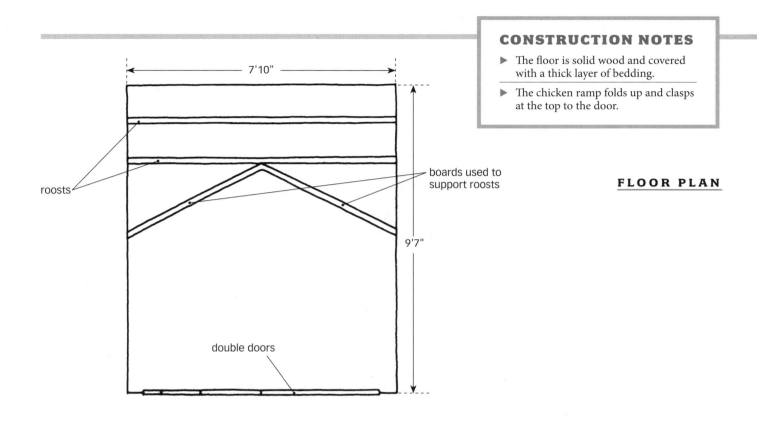

7'10"

roosts

boards used to
support roosts

9'7"

double doors

CONSTRUCTION NOTES

▶ The floor is solid wood and covered
with a thick layer of bedding.

▶ The chicken ramp folds up and clasps
at the top to the door.

FLOOR PLAN

FRONT ELEVATION

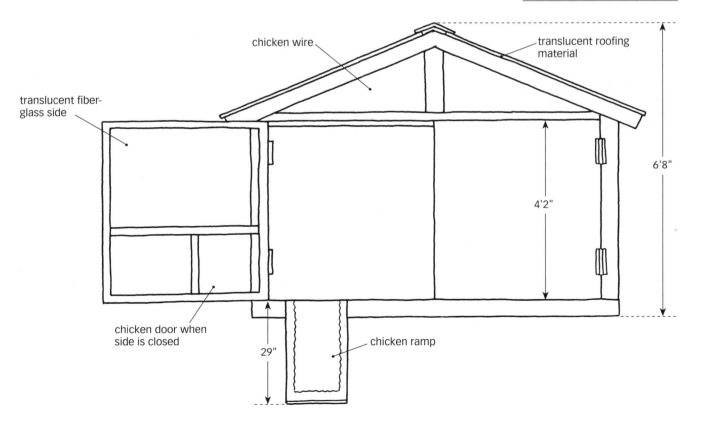

chicken wire

translucent roofing
material

translucent fiber-
glass side

6'8"

4'2"

chicken door when
side is closed

29"

chicken ramp

Sap Bush Eggmobile *(see color photo on page 154)*

The Sap Bush Eggmobile is built on the frame of an old barrel-style manure spreader. The floor is steel mesh so the droppings fall to the ground. The top is a single-pitched roof with chicken-wire inserts under the eaves. Two sets of 10-hole nesting boxes line one wall of the coop. Both sets can be accessed from the inside, and the one in the back can also be accessed from the outside. The people door is located on one end and the hens' door on the other. Jim and Adele keep a thick layer of bedding in the nest boxes. This encourages the hens to use all of the nest boxes instead of crowding into just a few. There are no roosts, and no bedding is on the floor. The Eggmobile is closed by 5:00 P.M. every day and opened in the morning. This is done to avoid developing broody hens and to discourage hens from piling up in the nest boxes. The result is clean eggs and less egg-processing time.

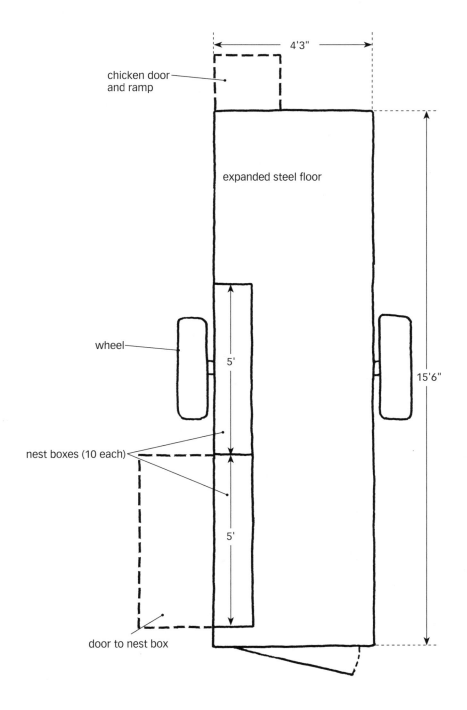

4'3"

chicken door and ramp

expanded steel floor

wheel

5'

nest boxes (10 each)

5'

15'6"

door to nest box

Sap Bush Laying-Hen Coops (continued)

FRONT ELEVATION

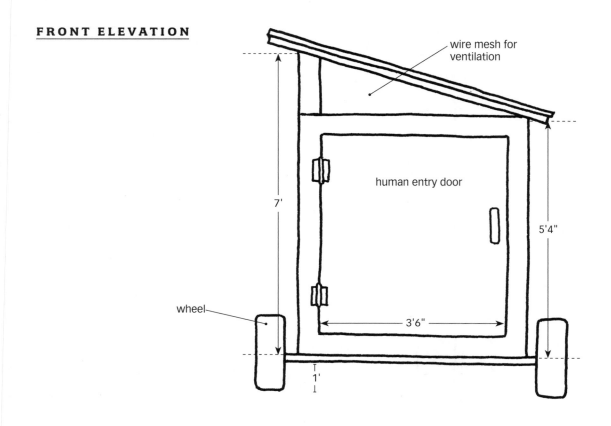

wire mesh for ventilation

human entry door

7'

5'4"

wheel

3'6"

1'

SIDE ELEVATION

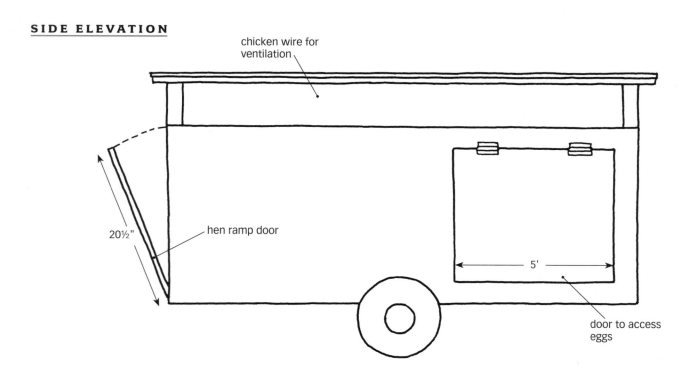

chicken wire for ventilation

hen ramp door

20½"

5'

door to access eggs

Sap Bush Brooder House

THE BROODER HOUSE at Sap Bush Hollow Farm in West Fulton, New York, is an old milk-can-cooler shed attached to the farm's historic barn. The shed was carefully renovated to provide adequate heat, ventilation, and predator protection. The ceiling and interior walls are insulated with roll insulation and covered in plywood. To provide plenty of light and ventilation, sliding windows are located about four feet from the ground and at intervals just off the ground. Window openings are covered with chicken wire and glass-paned sliding sashes. To eliminate drafts, secure openings with plywood inserts or sliding windows.

The brooder is heated with a propane pancake heater that hangs by a chain from the ceiling. The heater is hung just off the floor when the chicks are small and is raised as they grow. Two heat lamps are hung from the ceiling next to the propane heater and are used during the brooding of turkey poults.

For predator protection, openings are covered with chicken wire or hardware cloth. The concrete floor and the plywood-covered walls are checked often for signs that rats or other critters may be attempting to burrow into the brooder house. To prevent the chicks from suffocating should they pile up if cold or panicked, Jim and Adele collect the cardboard box lids that the chicks arrive in and staple the flat cardboard across the corners. The cardboard has airholes so that if the chicks pile up, the little ones can still breathe.

Water is supplied by a 5-gallon pail located on a raised shelf outside the coop. A water line goes through a hole in the wall down to an automatic bell waterer. A regular waterer is added for large batches of chicks. Old egg cartons and flats serve as feeders for the first week. The floor is covered in wood shavings.

The brooder house also serves as the winter house for the laying hens. A hen playpen was added off the side. Built from dog-pen panels, the pen is accessed by humans from an outside gate and by chickens from the indoor hen door and ramp. For predator protection, chicken wire covers the lower 2 feet of the dog run. A lightweight wood frame was built over the entire pen and is covered by a tarp. Once the chicks are big enough, they enjoy the pen on nice days, too.

SIDE ELEVATION

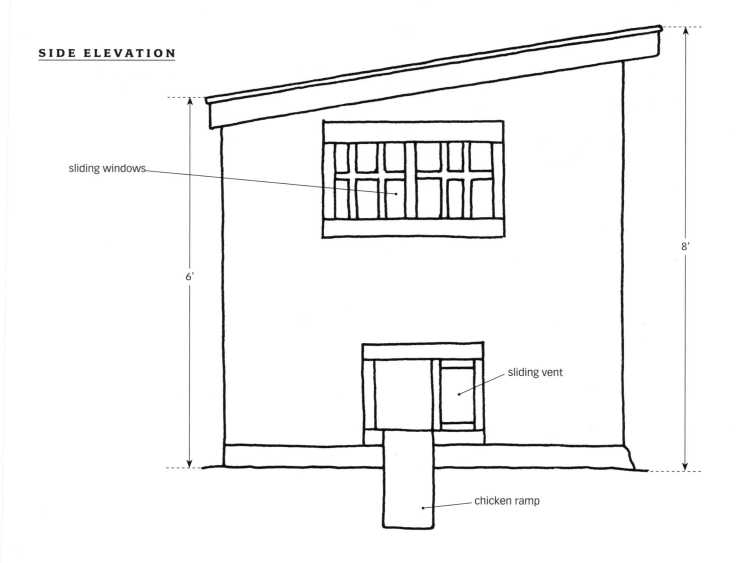

sliding windows

6'

8'

sliding vent

chicken ramp

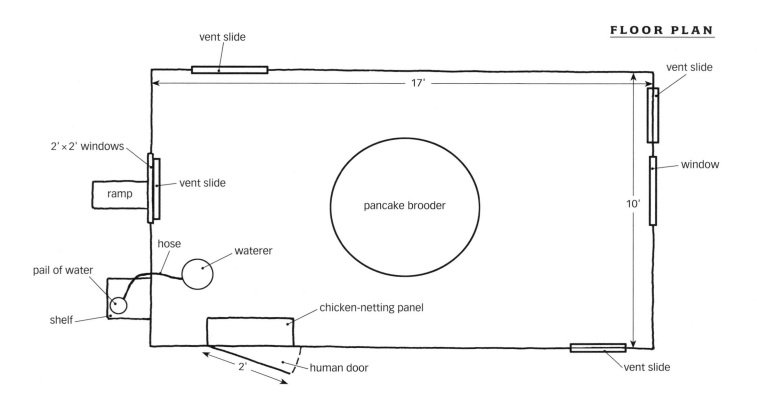

vent slide

vent slide

17'

2' × 2' windows

vent slide

window

ramp

vent slide

10'

hose

waterer

pail of water

pancake brooder

shelf

chicken-netting panel

2'

human door

vent slide

BROODER HEATER AND VENTILATION

The inside of the Sap Bush Brooder has a number of important features. The Hayeses use a propane-powered pancake heater. In addition, they use two protected lightbulbs, such as the one shown to the right, for extra light. Cardboard from the shipping container is tacked across the corner to prevent the chicks from suffocating if they pile up in the corners. The windows and hen door can be tightly closed to prevent drafts on chilly, windy days. The walls are insulated and covered with plywood to keep the coop warm and predators out.

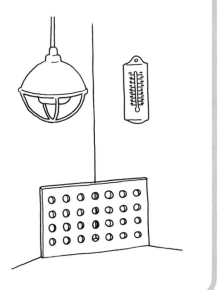

OUR FARM CSA

JENN WARD RUNS OUR FARM CSA, her family's fourth-generation farm in Easton, New York. Jenn follows organic methods in growing all of her vegetables and fruit. A few years ago, she added pastured poultry and laying hens to the operation to increase the organic matter in the farm's extremely well-drained and sandy soils and to increase farm revenues. Jenn direct-markets all of Our Farm CSA's produce at farmers' markets in the summer, at a winter farmers' market, and through the farm's community-supported agriculture (CSA) program.

In 2003, she attended a pastured-poultry training workshop and met Erika Marczak. After striking up a conversation, the two women discovered that Jenn needed help on the farm and Erika needed a farm on which to raise her pastured chickens. This began a unique opportunity for both farmers. Now Erika's work earns her room and board at the farm and a share of production for the weekly farmers' market she attends in the summer. Erika also raises broilers that she sells to local restaurants. In exchange, she works in the fields, helps run the CSA, and assists in managing the broilers and laying hens. The farm's overall production has increased through the efforts of both women, and each has helped develop the entire enterprise.

The newest additions to the CSA this year were pastured hogs, which were well received by the CSA members and by customers at the farmers' market. An Our Farm CSA membership also includes shares of the vegetable and berry production. In the spring, each member of the CSA pays for a share of the next season's produce. This provides Jenn with money for seed stock and other expenses for most of the farm's production. During the growing season, the produce is harvested and sorted into shares. CSA members come to the farm on a weekly basis to pick up their share, and some stay to participate in the farm operation. They are encouraged to help with the harvest, might be called upon to pitch in on chicken-processing days, and are an integral part of life on Our Farm. The CSA provides even production and cash flow for a portion of the farm's operation. This helps offset the variables of producing for farmers' markets, where sales are subject to dramatic fluctuations on a weekly basis.

Our Farm Broiler Pen

(see color photo on page 155)

HOLDS:
75 broilers

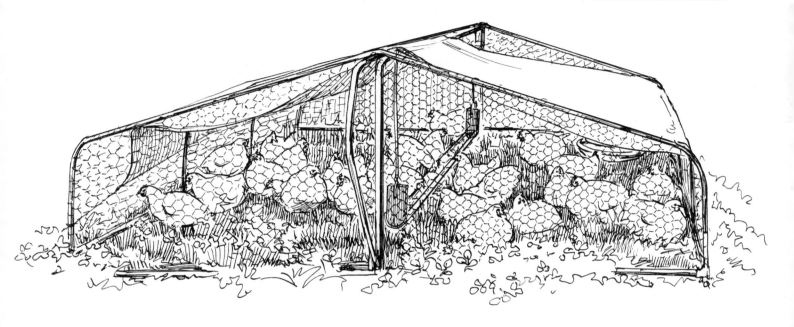

JENN WARD AND ERIKA MARCZAK raise 150 birds at a time on Our Farm CSA, an all-natural-vegetable farm in Easton, New York, about 30 miles north of Albany. Two lightweight metal broiler pens are moved through the cover crops in the gardens. To meet state processing requirements, Jenn's birds are processed on the farm and sold through the community-supported agriculture (CSA) program and at farmers' markets, and Erika's birds are processed at an off-farm facility and sold to local restaurants.

The broilers start out in the greenhouse chick brooder (see page 105) for a few weeks and then are moved to the broiler pens. The pens are partially covered with a light-colored tarp for shade and wind protection and to allow lots of light into the coop on sunny days. During the spring and fall, the pens are aligned side by side to increase wind protection. During the hottest part of the summer, the coops are staggered to increase airflow.

The metal frame was constructed with sections of EMT conduit cut and welded together. The walls are made of chicken wire attached to a metal frame with plastic ties spaced every few feet, and the coop has no floor. A feeder made with a PVC pipe cut in half, with 2×4 ends, is hung with rope or wire from the center support and runs almost the entire length of the pen. Water is provided by vacuum-style plastic waterers. Metal handles are attached to each end of the coop for easy moving. One top panel completely lifts off for easy access to the inside of the pen.

Predators can be stopped by surrounding the pens with electric netting, but Jenn uses a trusted guard dog to protect her broilers.

Our Farm Broiler Pen (continued)

SIDE PERSPECTIVE

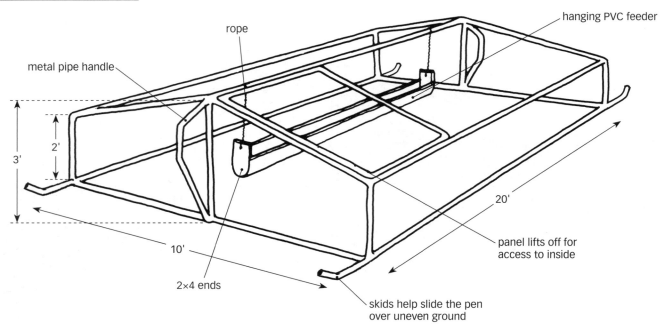

metal pipe handle

rope

hanging PVC feeder

3'

2'

20'

10'

2×4 ends

panel lifts off for access to inside

skids help slide the pen over uneven ground

OUR FARM PVC FEEDER

This inexpensive feeder is a PVC pipe that was cut in half lengthwise and cut to the length of the coop. Short 2×4 boards are screwed to the ends of the pipe. Ropes are attached to the top of the board ends, and the feeder is hung from the coop's center support. When the coop is moved, the feeder goes right along with it.

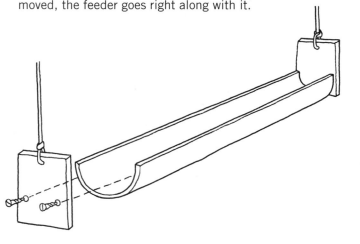

Soup Coop

HOLDS:
75 hens
20 nest boxes
6 roosts

MANY PASTURED-POULTRY producers consider hens spent after they have produced eggs for just a year or two. These hens are typically removed from the flock and either marketed as stewing hens or sold live to backyard producers for their flocks. On Our Farm CSA in Easton, New York, older hens are sent to the Soup Coop for one season before they are retired. The Soup Coop is a simple homemade hoop house built with 2×4s and 20-feet-long PVC-pipe hoops placed over stakes sunk into the ground. The hoop structure is covered with a tarp secured over the hoops, and a tarp is hung at each end to block sun and rain. Tomato stakes are nailed together to form roost platforms. These are placed across from two nesting-box units that stand back-to-back in the center of the coop.

The Soup Coop stays in the same spot for the entire summer. Electric netting fence is moved every couple of weeks around the coop throughout the season. After each summer, the ground beneath the pen is quite messy, so the following summer the coop is moved to another part of the field, and the fence is repositioned around it in the new section of the pasture. The Soup Coop hens may not be the most prolific producers on Our Farm CSA, but they are fertilizing a part of the farm that would otherwise not get very much attention, and their extra-large and jumbo eggs more than make up for their slower production.

Soup Coop (continued)

FLOOR PLAN

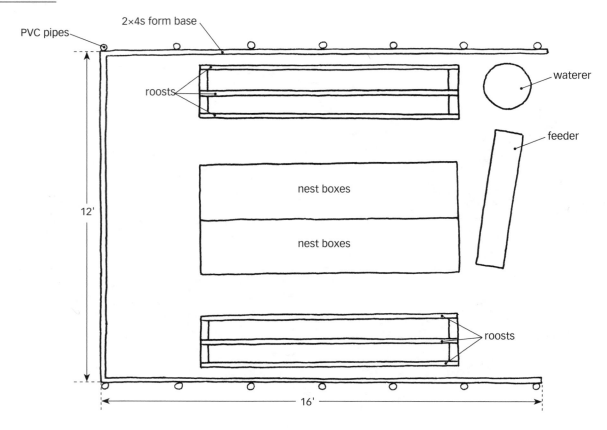

PVC pipes

2×4s form base

roosts

waterer

feeder

nest boxes

nest boxes

roosts

12'

16'

FRONT ELEVATION

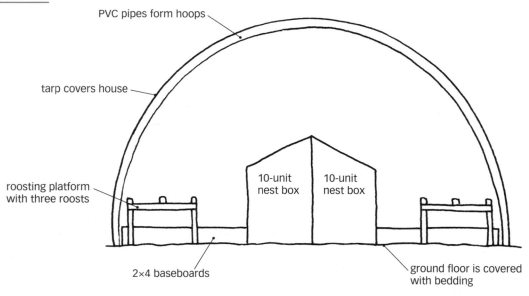

PVC pipes form hoops

tarp covers house

roosting platform
with three roosts

10-unit
nest box

10-unit
nest box

2×4 baseboards

ground floor is covered
with bedding

A-Frame Henhouse

(see color photo on page 153)

HOLDS:
150 hens
20 nest boxes
4 roosts

OUR FARM CSA'S 150 LAYING HENS spend their summers in a mobile A-frame coop. Built on a wagon frame, the A-frame is made with salvaged metal roofing. Nest boxes line both sides of the inside of the coop. The roosts are freestanding platforms made with old tomato stakes. Electric netting that is rarely charged surrounds the outside pen area. A 15-gallon pail inside the front of the henhouse feeds an automatic bell waterer hung from the front of the coop. Feeders are spread out on the ground behind the henhouse.

After the vegetables have been harvested, the gardens are seeded with cover crops, mostly clover. The hens are pastured on the fallow garden plots. After just a few years, the additional organic matter from the laying hens and broilers is helping the sandy soil retain water and nutrients. The quality and quantity of the vegetables increases each year, and the cover crops grow thicker and greener. The hens also produce eggs that are packed with nutrients and taste delicious because they've been allowed to pasture.

A-Frame Henhouse (continued)

FLOOR PLAN

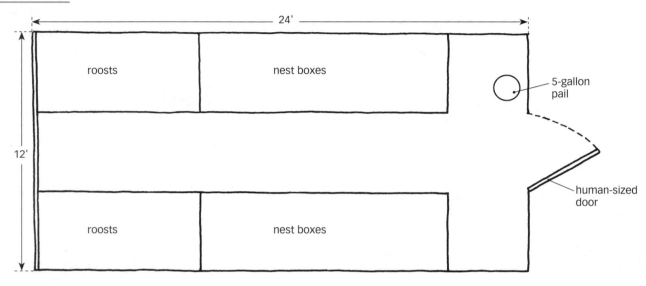

FRONT PERSPECTIVE

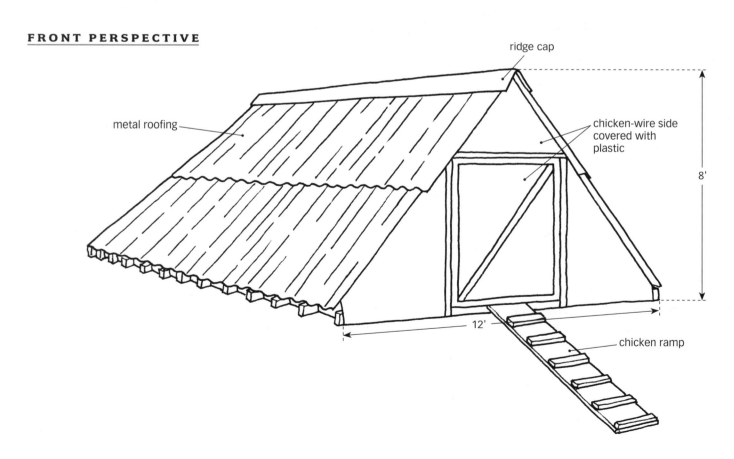

Chicken Greenhouse

HOLDS:
100 to 200 hens
20 nest boxes
10 roosts

"THE REAL VALUE of the chickens on Our Farm is what they do for the ground," Jenn Ward explains. Fellow farmer Erika Marczak continues, "We had our first tomatoes in June in the Chicken Greenhouse. The tomatoes just exploded out of the ground because of the nitrogen and phosphorus from the chicken droppings!"

Three vegetable greenhouses at Our Farm CSA in Easton, New York, serve as winter housing for the laying hens and as brooder houses for the chicks from spring to fall. Two hundred laying hens spend the winter in the 14-foot × 40-foot clear-plastic-covered Chicken Greenhouses. Jenn keeps adding shavings and aerating the bedding to create a very deep bedding pack. This composts in place and keeps the chickens warm. In the spring, the majority of the bedding is cleaned out and spread on the fields. During the summer, with the addition of two layers of 80 percent shade cloth, the same structures hold batches of 150 broiler chicks. Every three years, tomatoes replace chickens in the greenhouses.

In the spring, day-old chicks are started in the center portion of the greenhouse in old 20-bushel apple crates. Heat lamps are suspended in the center of the crates. When the chicks are a few weeks old, they are moved to pens within the hoop house. The pens are about 16 feet long and are surrounded by hog panels covered with chicken wire. Feeders are placed on opposite sides of the pen, and heat lamps are suspended from the hoops. The dirt floor is heavily bedded with wood shavings. Because the greenhouse can get extremely hot during the summer months, ventilation is provided by fans, an opening above the door, and rolling up the plastic sides and securing them with twine or bungee cords. On unbearably hot days, Jenn and Erika mist the birds with water.

The coop stays toasty warm during cold winter days. At night, the temperature drops quickly, and the inside of the coop quickly turns freezing. To solve this problem, Jenn and Erika are building 8- to 12-foot-long mini hoop houses to cover the roost areas inside the greenhouses. The mini hoops will be PVC or metal pipes covered in layers of plastic or tarps. Heat lamps will be hung from the hoops in the roost areas for use on the coldest nights.

FLOOR PLAN

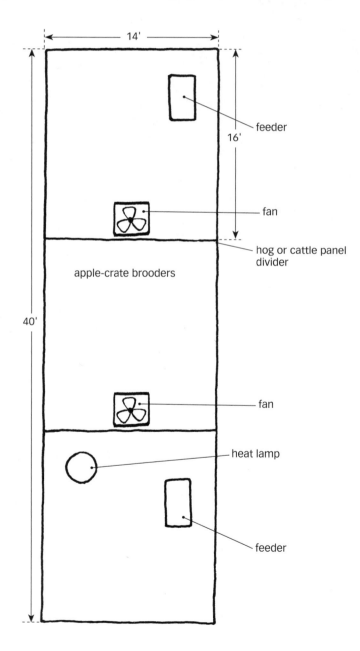

14'

16'

feeder

fan

hog or cattle panel divider

apple-crate brooders

40'

fan

heat lamp

feeder

CONSTRUCTION NOTES

▶ The configuration of the coop changes according to the season and farm needs.

▶ A shade cloth is used in addition to the plastic covering in the summer.

▶ The entire greenhouse is a chicken pen in the winter.

▶ During the summer, at least two pens house chicks and two to three large apple boxes serve as brooders.

SIDE ELEVATION

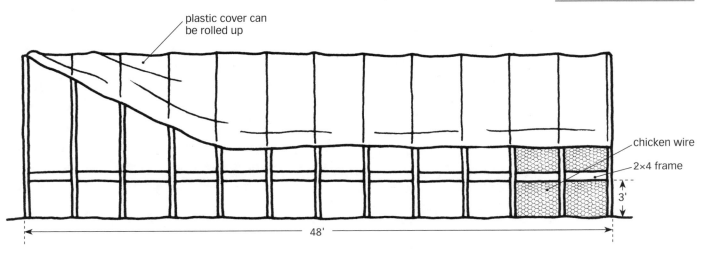

plastic cover can be rolled up

chicken wire

2×4 frame

3'

48'

SHILOH FARM AND RETREAT

AT THE SHILOH FARM and Retreat in Louisburg, North Carolina, Charles and Linda Gupton produce pasture-raised beef, pork, chicken, turkey, lamb, and eggs in addition to vegetables, fruit, honey, and flowers. When Charles, a professional photographer, and Linda, a writer, purchased the farm in 1999, they envisioned creating a sustainable-farm enterprise that would support a retreat center for people who needed a place to rebuild lives ravaged by abuse, addiction, or personal tragedy. Nonprofit status is being sought for the retreat center. When developed, retreat facilities will be available for small church-leadership meetings, marriage retreats, pastoral sabbaticals, and similar needs.

"Our vision for this farm also includes a strong desire to provide education and practical experience for others interested in sustainable agriculture careers. Through the retreat center, we hope to be able to provide housing and work for interns and apprentices who are interested in learning more about the farm and its operation," Charles and Linda explain on the farm's Web site (see page 162 for Web address).

The Guptons sell meat, eggs, vegetables, flowers and fruits through a prepaid preferred buyer's program and at local farmers' markets. The vegetables and fruits are grown without synthetic fertilizers and pesticides, and the meat is raised without antibiotics or growth stimulants. Techniques such as using composted plant and animal matter and cover crops help keep the soil fertile.

Chickens play an important role on the Shiloh Farm and Retreat and are critical to the operation's sustainability. The laying

Chickens play an important role . . . the laying hens and meat birds build the farm's soil fertility, while the pastured-poultry and egg sales provide revenue and attract customers to the farm.

hens and meat birds build the farm's soil fertility, while the pastured-poultry and egg sales provide revenue and attract customers to the farm. Shiloh boasts two mobile coops for the laying hens, as well as a meat-bird pen that can also house turkeys.

And the Guptons' vision of helping others to start agricultural careers is being realized. A beginning grass farmer has purchased one of Shiloh Farm's small mobile chicken coops so that he can start a small batch of pastured laying hens on his own farm.

Shiloh Mini Hen Retreat

(see color photo on page 152)

CHARLES AND LINDA GUPTON, owners of the Shiloh Farm and Retreat in Louisburg, North Carolina, designed and built their first mobile chicken coop to house 50 laying hens. The Mini Hen Retreat is a simple box-style coop with a single-pitched tin roof. The coop is built on a single-axle trailer frame with small rubber tires to make it easier to rotate the coop around the pasture. Hinged cutouts across the back provide access to the nesting boxes from the outside. A larger hinged cutout beneath the nesting boxes provides room for feed storage. The chickens come and go through hinged doors located at the sides of the coop. Roosts are made from two branches that were cut from nearby woods and attached to the inside walls of the coops. Covered feeders and waterers are located directly on the ground outside of the coop.

Shiloh Mini Hen Retreat (continued)

FLOOR PLAN

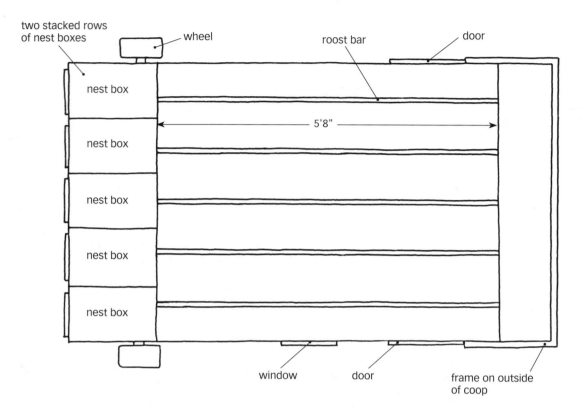

two stacked rows of nest boxes

wheel

roost bar

door

nest box

nest box

nest box

nest box

nest box

5'8"

window

door

frame on outside of coop

REAR ELEVATION

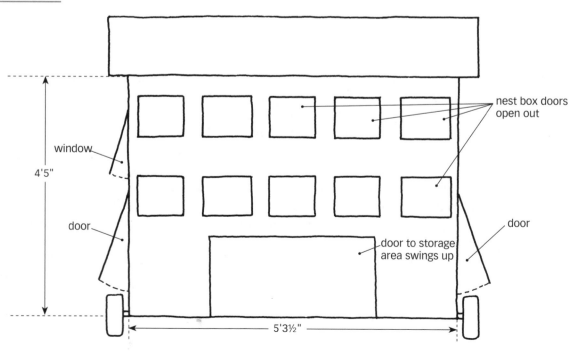

nest box doors open out

window

4'5"

door

door

door to storage area swings up

5'3½"

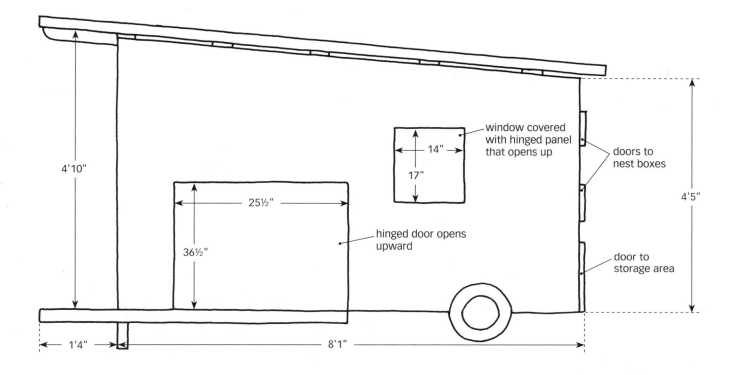

window covered with hinged panel that opens up

14"

17"

doors to nest boxes

4'5"

4'10"

25½"

36½"

hinged door opens upward

door to storage area

1'4"

8'1"

Shiloh Hen Retreat

WHEN DEMAND FOR the pasture-raised eggs on Shiloh Farm and Retreat in Louisburg, North Carolina, quickly exceeded supply, Charles and Linda Gupton built the Hen Retreat to house one hundred laying hens. This coop is a larger version of the Mini Hen Retreat with a few modifications: it has a wide overhang around the sides of its single-pitched roof, it features full-sized doors and windows, and it has an area on the side (the end with the hitch) for a covered water barrel and feed-storage bins. Waterers and covered metal feeders are placed on the ground outside of the coop.

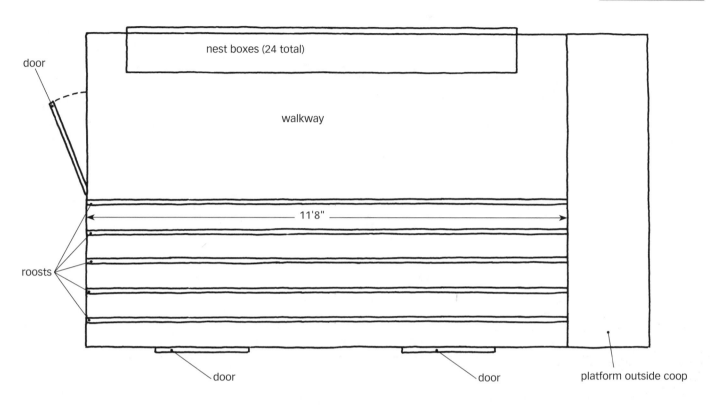

door

nest boxes (24 total)

walkway

roosts

11'8"

door

door

platform outside coop

GRAVITY WATER SYSTEM

Water is fed by gravity from a large water barrel mounted on the side of the Shiloh Farm Hen Retreat. A float attached to the water barrel allows the barrel to fill as needed. A black barrel works best because it absorbs heat in the winter, which keeps the water from freezing, and blocks sunlight in the summer, which prevents algae from growing inside.

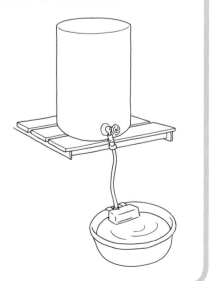

SIDE ELEVATION

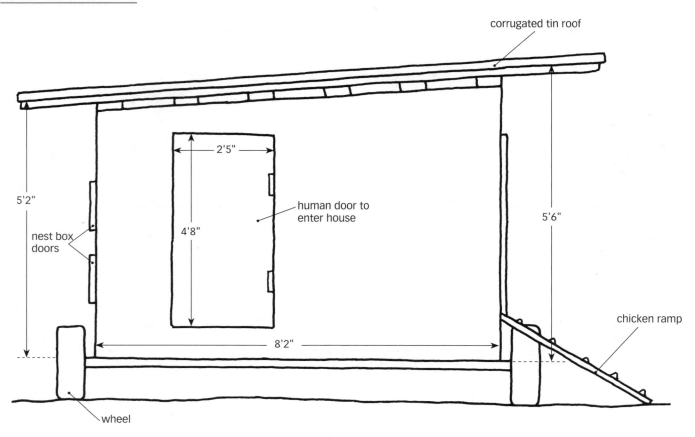

corrugated tin roof

2'5"

human door to
enter house

5'2"

4'8"

nest box
doors

5'6"

chicken ramp

8'2"

wheel

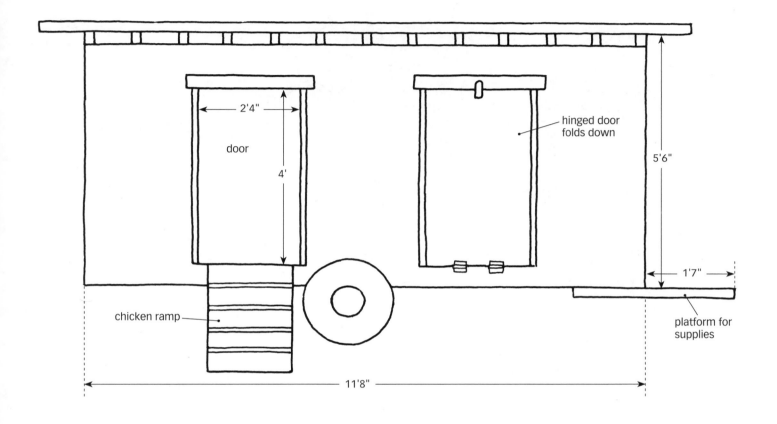

2'4"

door

4'

hinged door
folds down

5'6"

1'7"

chicken ramp

platform for
supplies

11'8"

Shiloh Broiler Retreat

HOLDS:
75 broilers

THE BROILER RETREAT on Shiloh Farm is an example of a lightweight and easy-to-move broiler or turkey pen that is designed to handle the extreme heat of a typical North Carolina summer. The pen is built using salvaged aluminum electrical conduit pipe and chicken wire. A tarp is cranked over the top to provide sun and rain protection when needed. The sides can be covered with empty feed bags to block the sun, wind, or rain. Two separate layers of poultry netting, set up about 2 feet apart, provide predator protection. Because the chickens will roost on the ground against the fence at night, the inside fence is not charged. The outside fence is charged with a solar charger.

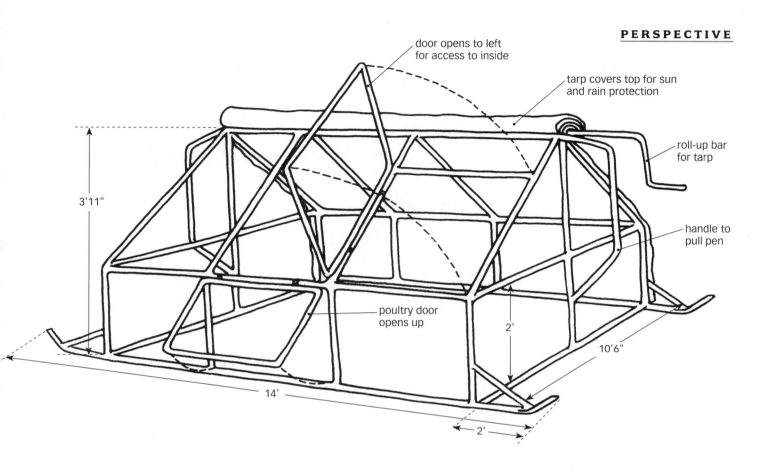

door opens to left
for access to inside

tarp covers top for sun
and rain protection

roll-up bar
for tarp

handle to
pull pen

3'11"

poultry door
opens up

2'

10'6"

14'

2'

HOUSING BROILERS VS. LAYING HENS

Broilers are much heavier birds than laying hens, and their legs are often not very strong. To prevent injuries, mobile broiler pens typically sit directly on the ground, and roosts are not provided. Broiler pens that are located on the ground and moved by hand must be lightweight. As a result, much care should be taken to make sure that these types of broiler pens have all the minimum requirements (shelter, feed, water, ventilation, and predator protection) while still being lightweight enough to move easily by hand. Laying hens can move about much more readily and are able to climb ramps, so their pens can be built on trailer frames designed to be moved by vehicles. This allows the coop owner to add more features if desired. Nest boxes, roosts, and inside feed storage are the main accoutrements that are likely to be added to a mobile laying-hen coop.

Cotton Trailer Coop

HOLDS:
250 hens
50 nest boxes
12 roosts

WHEN DEMAND EXCEEDED the capacity of the Hen Retreat (see page 112), Charles and Linda Gupton of Shiloh Farm and Retreat in Louisburg, North Carolina, converted an old cotton trailer that had been left on the farm when they purchased it. The Cotton Trailer Coop is designed to house more than 250 laying hens and is used in addition to the Hen Retreat. As a result, the Mini Hen Retreat (see page 109) was retired.

Using the knowledge they had gained from the first two coops, especially the problems inherent with a single-pitched roof on such a large coop, the Guptons designed the Cotton Trailer Coop to have a gable roof. This double-pitched roof makes the coop more stable on windy days and protects both sides

of the coop when it is raining. To provide adequate ventilation in southern summers, one side of the coop is open and covered with chicken wire. The other side consists of hinged doors that allow access to numerous roll-out nesting boxes (see box on page 121) that were salvaged from an old conventional chicken house. The bottoms of the original nesting boxes were tilted slightly from the front to the back and had very small openings that allowed the eggs to roll out onto conveyor belts in the conventional system. Charles enlarged the openings so that the eggs could be accessed from the outside of the coop while still being able to roll out for ease of egg collection. The nesting boxes are lined with rubber mats, which can be easily removed to facilitate cleaning. The V-shaped structures are 2×4s that are attached to the floor and ceiling. The crossbars that sit between these 2×4s are the roosts.

The coop's wood-slat floor was also salvaged from the conventional chicken house. The Guptons prefer this type of floor over a chicken-wire floor because the chicken droppings easily fall to the ground, and the smooth wooden slats are softer on the hens' feet. Two rain barrels attached to a platform over the wagon tongue gravity-feed water to waterers placed on the ground. Feed is contained in storage areas inside the back of the coop. Old chicken feeders salvaged from the conventional chicken house are placed on the ground behind the coop. Predators are kept at bay with poultry netting and a solar charger. The coop is moved to new pastures at least once a week.

CONSTRUCTION NOTES

► There is 83 feet of roost space.

► To build the roosts, construct two upside-down A-frames out of 2×4s; cut out three notches, evenly spaced along each of the 2×4s, and rest the roosts inside these.

FLOOR PLAN

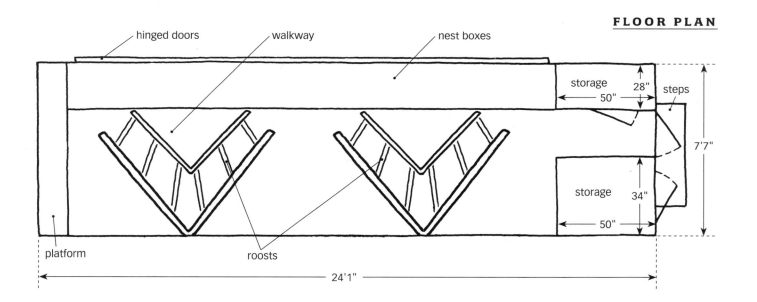

hinged doors   walkway   nest boxes

storage 50" 28" steps

7'7"

storage 34" 50"

platform   roosts

24'1"

Cotton Trailer Coop (continued)

SIDE ELEVATION

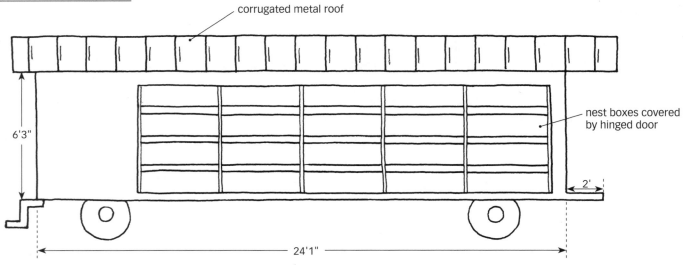

corrugated metal roof

nest boxes covered by hinged door

6'3"

2'

24'1"

REAR ELEVATION

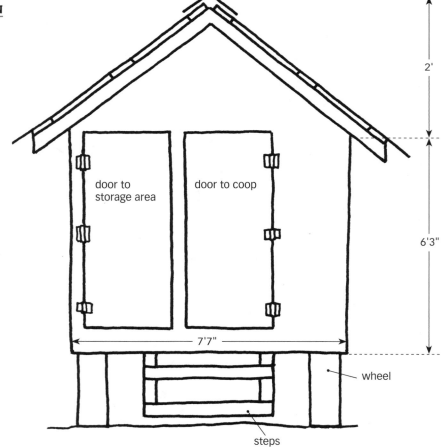

2'

6'3"

door to storage area

door to coop

7'7"

wheel

steps

NEST BOXES

Multiunit metal nest boxes can be found new and used from a variety of sources, including farm stores, old poultry houses and farms, and farm-equipment auctions. They can be very easily incorporated in a coop design and can allow for inside and outside access. The bottoms can be the original metal inserts, or they can be replaced with other materials such as contractors' fabric (a stiff metal mesh), plywood, or some other material.

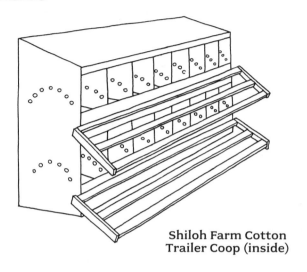

Shiloh Farm Cotton Trailer Coop (inside)

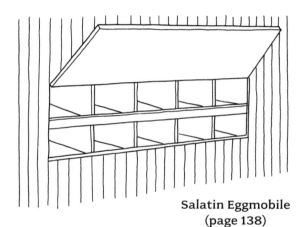

Salatin Eggmobile (page 138)

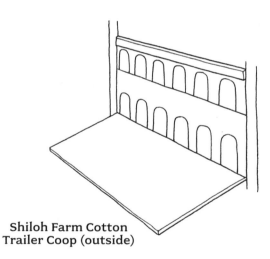

Shiloh Farm Cotton Trailer Coop (outside)

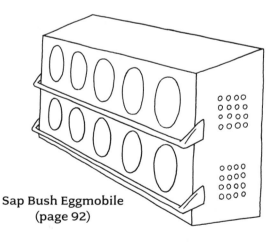

Sap Bush Eggmobile (page 92)

SWEET TREE FARM

FOR THE PAST SIX YEARS my husband, Frank Johnson, and I have been converting our 200-acre former crop farm in Carlisle, New York, into a grass farm where we produce beef, pork, eggs, and maple syrup. We intensively graze the beef, which means that they are moved to new small paddocks every other day. This way, they get the best greens in the pastures. And in turn, the pastures are naturally fertilized, aerated, "mowed," and then rested until they are ready for the next pasture rotation. We market all of our meat, eggs, and maple syrup directly to customers from the farm and at farmers' markets.

When we started grass farming six years ago, Frank had some experience with raising chickens, but I knew absolutely nothing about them. Instead of starting with chicks, we purchased 15 pullets from a neighbor. I quickly became accustomed to the birds and their habits and soon found myself pulling a lawn chair into their pen so the "girls" and I could chat. Our 15 birds grew to 75, then to 100, until we reached our current level of 200 hens. The hens are a variety of heavy-breed layers chosen for their ability to adapt to our pasture system and for the multicolored rainbow eggs that they produce. We currently have Rhode Island Reds, Barred Rocks, White Rocks, Black Australorps, and Buff Orpingtons, as well as Araucanas for their green and blue eggs.

We move our beef herd to fresh pastures every other day and then follow about four days behind them with the mobile chicken coop. This four-day delay allows the bugs to hatch in the cow pies. When the bugs are hatching, the chickens scratch through each cow pie, searching for bugs. As they scratch and eat the bugs, the chickens drive the manure directly into the ground, instantly depositing organic matter in the soil. If we move the chickens into the paddock too early, before the bugs begin to hatch, the hens will not attack the cow pies as aggressively. If we move the chickens too late, the bugs have hatched and flown away, and the cow pies are much less attractive to the hens. It is crucial that the chickens get to as many of these manure piles as possible. If a cow pie is left intact, it takes a year or more for it to break down into the soil. The cows simply refuse to eat the grass under or around a cow pie until the pie has completely broken down. If a chicken scratches a cow pie into the ground, however, the cows will eat the grass in that area in the very next paddock rotation. The beef/chicken cycle is critical to optimizing the soil fertility and grass palatability on our farm.

We collect eggs often and refrigerate them immediately. The occasional cracked or dirty egg is tossed to the pigs or the dogs. Most of our eggs we sell at the Saturday farmers' markets. On Sweet Tree Farm, we do not use artificial light to increase egg production during the winter. As the number of hours of natural light decreases in the winter, egg production decreases. Providing supplemental light (up to 7 hours) for a total of 15 to 16 hours of light per day would force the hens to lay more eggs. We avoid this practice to keep our production as natural as possible. The hens sure like it, and we get a reprieve from processing eggs in the middle of the winter.

Sweet Tree Birthday Coop

YEARS AGO, my father built a small garden shed that my parents gave me for my birthday. I used the shed all of the time and moved it a few times as I progressed from an apartment to a house to our farm. Time and the moves took a toll on the shed, and it eventually ended up behind the barn, neglected but not quite forgotten.

When we decided to add chickens to our agricultural enterprise, we searched for a coop design that would fit the farm. Our place, located in Carlisle, New York, has 200 acres of hills, flat terrain, and a few rough old farm lanes. The coop had to be mobile and sturdy enough to be pulled with a tractor over rough terrain for fairly long distances (long at least to a chicken). After looking at many coop styles, we remembered the old garden shed behind the barn. After we con-

verted the shed into a mobile henhouse for 50 chickens, my parents were very surprised, to say the least, to see their birthday present turned into one of the few mobile chicken coops in our county!

To convert the garden shed into a henhouse, we bought an old single-axle trailer frame from a neighbor. Because the shed did not have a floor, Frank used rough-cut 4×4 lumber from a nearby sawmill to build the frame for the floor directly on the trailer. He used the tractor bucket to pull the coop up wooden ramps onto the trailer base. Then he attached the coop to the trailer floor with lag bolts. The floor is framed with rough-cut boards and covered in chicken wire so the chicken manure can drop to the ground and fertilize the fields as the coop is moved. In cold weather, fitted boards are screwed over the chicken-wire portions of the floor to block drafts. During the summer, the floorboards are stored by screwing them to the inside ceiling of the coop.

For ventilation, window holes were cut out of the sides and covered with chicken wire. The wood siding that was cut out for the windows was fashioned into hinged shutters that we can close when the weather is particularly nasty. One side of the wall was cut out, and a 10-hole nesting box was hung on the inside. On the outside, we added a shutter with hinges across the top. The shutter is propped up with a stick when we collect eggs from the outside of the coop. Five roosts are located across the inside back of the coop and are made from sturdy sticks found in our maple woods. Two galvanized chicken feeders are hung from nails on the outside of the coop. We store feed in a heavy-duty plastic garbage can, using bungee cords to hold the can in place on the wagon tongue.

The chickens are watered from an open rubber pan and a 5-gallon plastic vacuum-style chicken waterer. All of our pastures are connected to an aboveground plastic water line that is supplied by the farm's drilled well. Water valves are located at regular intervals along the line so that each paddock has access to water. Because the chickens follow the beef through the paddocks, water is available nearby no matter where the chickens are on the farm. We manually fill the chicken waterers and double-check them during chores. Predators are kept at bay with two sections of portable electric sheep netting fence powered with a portable charger that runs on D batteries. We prefer the portable electric sheep netting fence because it is lighter and easier to move than the portable electric poultry netting fence. The hens can pop through the larger holes in the sheep netting, but they quickly learn that it is safer inside the pen than outside of it.

FLOOR PLAN

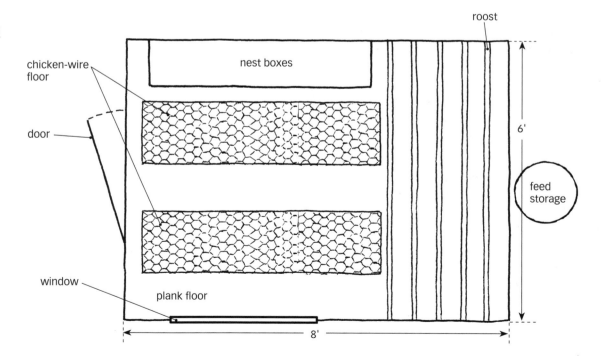

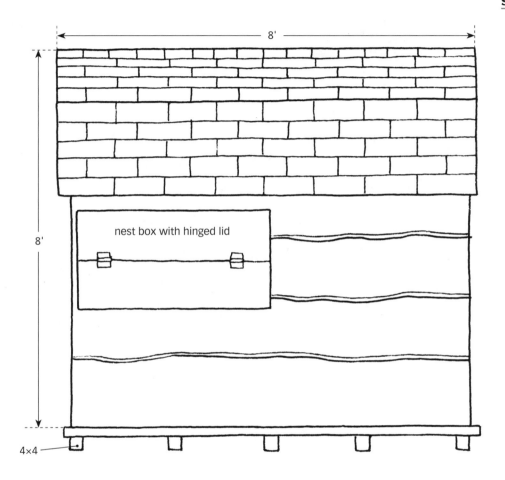

nest box with hinged lid

8'

8'

4×4

Coop de Grass

HOLDS:
200 hens
18 nest boxes
8 roosts

THE DEMAND FOR the pastured eggs from our farm in Carlisle, New York, grew faster than we had expected, so we purchased more chicks. When our first coop proved too small, we designed and built the larger Coop de Grass. The coop turned out so nice that we thought about using it as a mobile cabin at the fishing pond, but the demands of the flock took precedence.

The coop meets all of the hens' needs: it has ample space for them to roam around and has plenty of nesting boxes and roosts. The coop is constructed on a double-axle frame. Frank built portable ladders that hang on the outside of the coop so that the kids and I can reach the top row of nesting boxes from the outside. The floor is chicken wire.

The roosts are stacked like stairs, starting near the floor by the nest boxes and ending near the tops of the windows. The nest boxes are homemade wooden boxes with purchased plastic inserts in the bottom. We installed hinged doors on the outside of the coop so that we can collect the eggs from the outside. The plastic inserts are roll-out trays. They didn't work as well as we had hoped, though (the hens' feet get caught, and the eggs roll out either too fast or not at all), so this winter we will replace them with flat-bottomed trays.

Because of its single-pitched roof, the coop is most properly suited to flat fields or areas where there is wind protection. When we occasionally move it to a sidehill or are expecting severe winds, we prop long 2×4 T-bars, from the ground to under the eaves on both sides, to give the coop the added stability it needs.

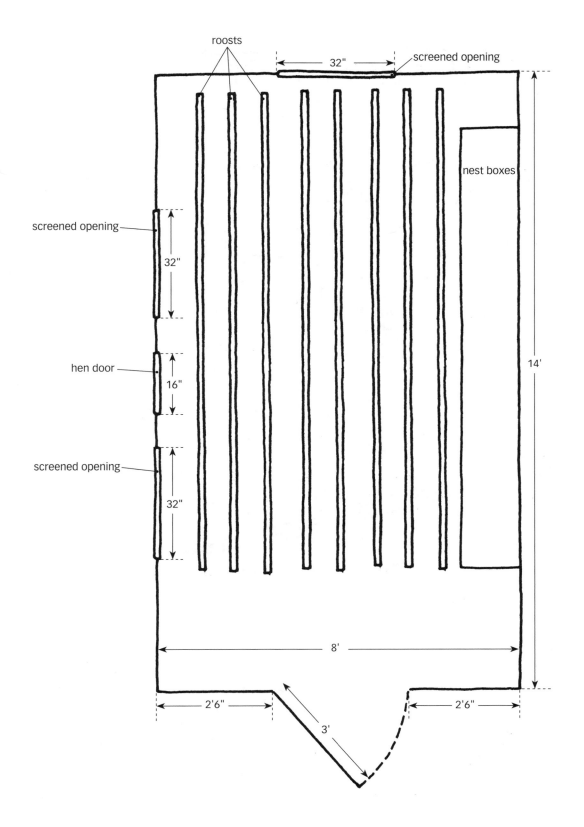

roosts

32"

screened opening

screened opening

32"

hen door

16"

screened opening

32"

nest boxes

14'

8'

2'6"

3'

2'6"

Coop de Grass (continued)

FRONT ELEVATION

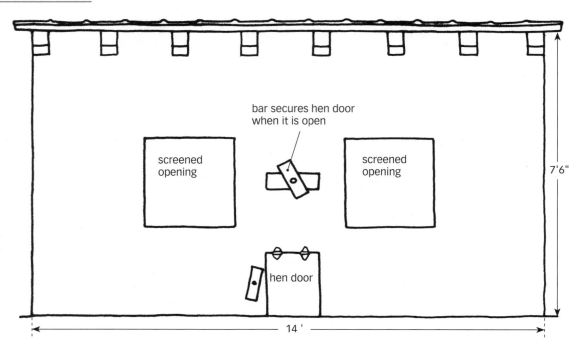

bar secures hen door
when it is open

screened
opening

screened
opening

7'6"

hen door

14'

SIDE ELEVATION

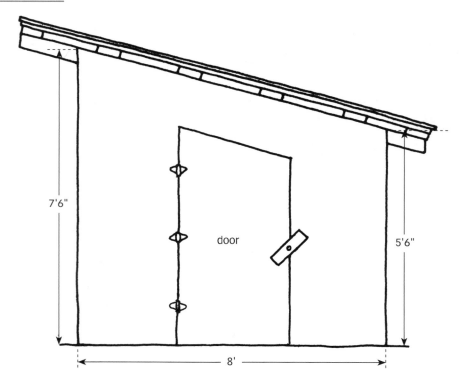

7'6"

door

5'6"

8'

Winter Coop

HOLDS:
200 hens
20 nest boxes
10 roosts

OUR UPSTATE NEW YORK winters are tough for humans and chickens alike. Temperatures drop below zero for days and weeks at a time, and winter winds continually blast across the fields. We have either several feet of snow or bare, icy ground from mid-December through March. We have had severe snowstorms in early October and in late May. Our harsh winters mean that our mobile coops are summer facilities, and that more-secure winter facilities are needed.

If we had a closed shed large enough to hold the mobile coops, we would simply back the coops into the barn, set up a chicken fence for the winter, and not worry about separate winter facilities. Unfortunately, we do not have such a barn available. We do have an old permanent chicken coop circa 1930. The first year we needed to house chickens for the winter, the old coop was full of junk left from many previous owners, so we did not use it. Instead, we built a coop inside part of the old dairy barn using 2×4s and chicken wire, and this sufficed for a time.

The next year, we had doubled production and needed more space. This time, we took another look at the old chicken coop. Once we had cleaned it out and began using it, we realized the value of this old-style coop and the chicken know-how that went into the design. The coop is perfectly positioned so that the long, nearly windowless back side and narrow, windowless end can block fierce winter winds that come from the north and west, while the east-facing windows along the long front soak up winter sunlight.

On warm spring days, the bank of coop windows across the front and the two smaller windows at the back provide cross ventilation to take advantage of spring breezes. We discovered three small openings tucked beneath the front soffits. The openings have sliding doors that can be manually adjusted to provide winter ventilation when the windows are tightly closed. A larger vent in the wall above the front windows provides even more control of ventilation in the winter. The coop has chicken-sized access doors on both the front and the back so that the chickens can freely range on both sides of the coop.

The coop has two areas separated by a wall and a door. Each section has galvanized feeders hanging from the ceiling and rubber water tubs on the ground. Five levels of roosts approximately 8 feet long are located in each section. A 10-unit nesting box is attached to each side of the wall that separates the sections. When we have a large batch of hens that are close in age, we keep the door open and the hens have access to the entire coop. If we are starting a new batch of pullets, we prefer to keep the younger, higher-producing hens separate from the older ones. We keep the pullets in one section of the coop and the older hens in the other. In the spring, the older hens are pastured in our smaller mobile coop. They provide bug control around the farm buildings and in the pastures near the barns. The younger hens are pastured in the larger mobile coop and follow the cows through the pastures.

Instead of constantly cleaning out the chicken coop over the winter, we use a bedding-pack system to keep the hens warm and comfortable. We layer wood shavings and straw in the coop every few days. Over the course of the winter, the bedding pack grows to a foot or two in depth. Mixed with the chicken manure, the shavings and straw heat slightly and begin to decompose into organic matter. The coop stays clean and smells fresh, the chickens stay warm and dry, and in the spring we have lightweight and high-quality organics to add to our compost pile, share with friends for their gardens, or spread directly on the fields.

WINTER COOP VENTS

The vents in the Winter Coop are very simple rectangular openings with a sliding board cover. They are set in the soffits on the outside of the coop. We added chicken wire to cover the openings on the inside of the coop. This type of vent could also be installed directly in the sidewall, but it should be located high enough so that the birds are not subjected to drafts. These vents would also work well in a brooder building, but they must be set about a foot off the ground so the chicks get the benefit of the fresh air without being chilled by drafts.

FLOOR PLAN

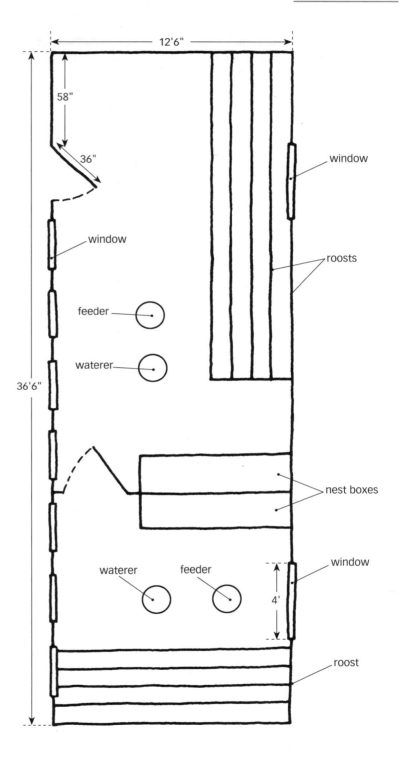

Winter Coop (continued)

FRONT ELEVATION

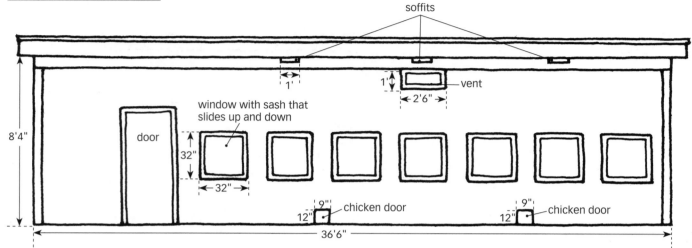

OLD POULTRY EQUIPMENT ON FARMS

Over the years, as chickens disappeared from small family farms, many chicken coops were simply closed with all their equipment still intact, or the poultry equipment was piled away in a corner of another farm building and forgotten. If a farm has not been sold and emptied and there are still old barns and chicken coops standing, you just may be in luck. Most farmers enjoy helping new producers get started, so ask the farmers in your area if they have any old chicken equipment around their farms. Look for brooders, brooder heaters, old nest boxes, feeders, waterers, roosts, and other equipment. Chances are, you will scrounge up quite a bit of equipment at very little to no cost.

If you come across old chicken coops, ask for permission to study them for tips on coop layout, ventilation, window design and placement, location of hen doors, and predator protection. These old coops were designed by people who really knew chickens, and the coops work well in all seasons. Old coops are fairly easy to identify — most are long, one-story buildings with windows across at least one side, and they will have vents and hen doors located around them.

POLYFACE FARM

MY FAMILY AND I visited Joel Salatin's Polyface Farm during an August heat wave and drought. The Shenandoah Valley farm is located a tad off the beaten path (the farm's customers will laugh at the understatement!), and as we tried to find the farm, we passed field after field filled with Angus cows and calves knee-deep in dried brown grass. We knew we had found Joel's farm when the landscape changed in an instant from brown to deep, vibrant green. This was the greenest grass we had seen — except for golf-course greens — since leaving Raleigh, North Carolina, earlier that morning. My husband and I had read all of Joel's books and had attended a workshop a few years ago in Albany, New York, in which Joel presented a slide show of his farm. Even though we were familiar with his grazing practices and thought we knew what to expect, we were still impressed with the exceptional condition of his pastures.

Pioneering grass farmers Joel and Teresa Salatin — together with their daughter, Rachel, and son, Daniel, and his family — operate this diversified grass farm located at the base of the Appalachian Mountains near Swoope, Virginia. The Salatins produce and direct-market pastured poultry, turkey, beef, pork, eggs, and meat rabbits, and their business is booming. "We did fifteen thousand broilers this season," Joel says. "We have seven hundred turkeys in the turkey brooder that will be ready for our Thanksgiving sales."

Joel and his family have greatly contributed to the growing pastured-poultry and sustainable-agriculture movements with their tireless efforts to educate fellow farmers and the consumer about the benefits of pastured poultry. Joel coined the popular grass-farming phrases "egg-mobile," "salad bar beef," and "pigaerator pork," among others, in his books (see pages 162—163 for a list of the books). Joel and his farm have been featured on national television. He often contributes to the *Stockman Grass Farmer* and other magazines, and Joel and Daniel are popular speakers at grass-farming conferences.

The Salatins are also pioneers in direct-marketing their grass-fed meat and poultry. Broilers, turkeys, eggs, and all of the farm's products are sold by order direct from the farm, at local farmers' markets, and to restaurants in nearby cities. Joel's newest marketing model is a series of food-buying clubs consisting of urban customers who purchase large quantities of products in a single order. Joel arranges the delivery to a pickup point in their vicinity. "The whole drop point has a minimum-average-volume requirement," says Joel, "but individual patrons can get as much or as little as they want." The cost of handling and delivery is added to the order. The food-buying club distributes the products to individual members. Given its success, Joel is convinced that this direct-marketing method is the wave of the future.

Salatin Turkey Coop

SOMETIMES A COOP can be as simple as a frame and a roof. This lightweight coop on Joel Salatin's farm near Swoope, Virginia, is easy to move yet sturdy enough to stand up to any abuse the turkeys can give it. The open-framed coop is built with metal pipes. The roof supports extend beyond the frame on both sides for more shade protection. The roof tarp can be rolled up if it is not needed for shade or protection from the weather.

Wheels are attached at each corner of the frame. Only two of the wheels swivel, so the coop can turn only on one end. The coop is heavy enough to prevent the wind or the turkeys from being able to push it around, but it can still be moved easily to a new area of the pasture. A box mounted on the back corner holds grit for the turkeys. To provide protection against predators, the pen is surrounded by electric poultry netting.

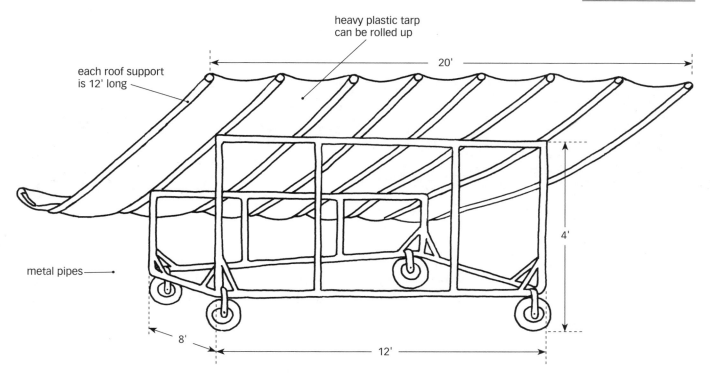

heavy plastic tarp
can be rolled up

20'

each roof support
is 12' long

4'

metal pipes

8'

12'

A NEARLY PERFECT SYSTEM

Polyface Farm is a diversified ecosystem in which every animal serves multiple purposes. The poultry are rotated often so they always have access to fresh pastures and so they can instantly inject organics into the soil; no waiting for that compost pile to turn in this system! Although the beef animals instinctively will not eat the grass that surrounds an untouched cow pie, they will eat the grass that regrows where the laying hens have scratched the cow pie into the ground.

The chickens' natural diet produces meat and eggs loaded with vitamins and minerals. These high-quality chickens, turkeys, and eggs command a good price and, once introduced to the consumer, create increased demand through word of mouth and repeat customers. It is very nearly a perfect system for the poultry and eggs, the cows, the farm, the farmer, and the consumer.

Salatin Portable Broiler Pen

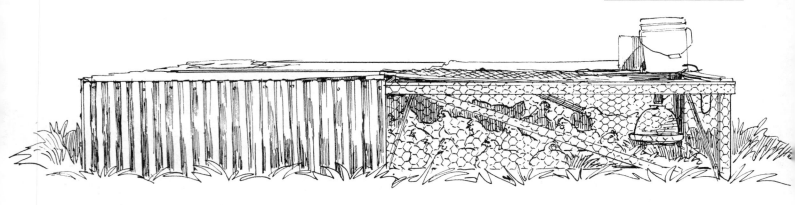

EVERYTHING ABOUT Joel Salatin's grass farm is impressive, and the broiler-production system is no exception. On a hill overlooking the farm buildings, more than 30 broiler pens zigzag across the hill as far as you can see. Each pen holds 80 to 90 birds, and Joel will produce 15,000 birds in a season. This system is successfully emulated in small to large pastured-poultry operations throughout the country.

The floorless pens have a very simple design and can be made with salvaged materials. Each is 10 feet × 12 feet × 2 feet. The back half has aluminum sides, and the other half is enclosed in chicken wire for ventilation and predator protection. The roof is also aluminum, with an open section covered with chicken wire for ventilation. Because of the heavy weight of the broiler chickens, roosts are not safe and are not provided. Five-gallon pails are placed on top of the pens, and water is fed by gravity from the pails to automatic bell waterers hung inside the pens. The pails are filled with water pumped from a nearby pond. Bulk feed is stored in large used fuel tanks that are cut 4 feet high, scrubbed with a wire brush, left in the rain for at least a month, and dried before use.

When the coops are moved, the empty feed tanks are tipped over and rolled to a new spot near the coops. The coops are moved with a dolly designed by Joel's brother, Art, and modified by Joel. A 4-foot-square handle on a bent axle made of ½-inch pipe, this dolly allows one person to move a pen in about 30 seconds. The axle is 2 to 3 inches above the ground on the ends and about 6 inches above the ground in the middle. Being bent, it provides better ground clearance than a straight axle. Two prongs located near the wheels slide under the edge of the pen. The meat pens are pulled to fresh grass at least once a day. The birds walk on the grass as the coop is moved to new pasture.

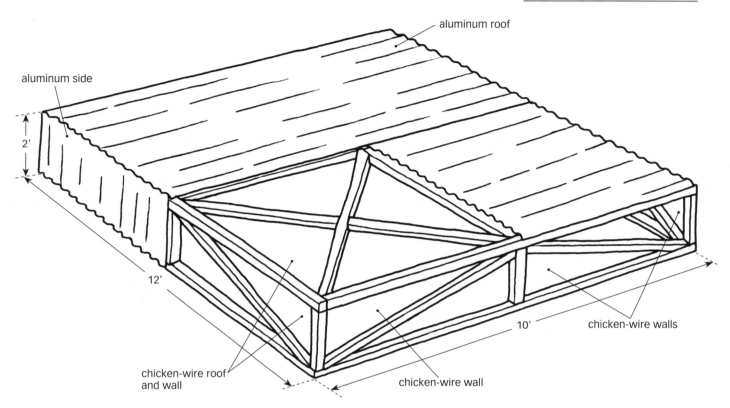

aluminum roof

aluminum side

2'

12'

10'

chicken-wire walls

chicken-wire roof and wall

chicken-wire wall

PROCESSING BROILERS

Where regulations allow, many broiler producers prefer processing their birds on the farm and marketing the 4- to 6-pound roasters directly to the consumer. On-farm processing facilities typically include an area where the birds are humanely butchered and a separate area where the carcasses are cleaned, plucked in a machine, bagged, and chilled.

This method is preferred because it enables the grower to have control over every aspect of production, processing, and marketing. Where regulations do not allow on-farm processing, birds must be processed at federal- or state-inspected facilities. Broiler producers must learn and keep current with the regulations that apply to their operation.

Salatin Eggmobile

HOLDS:
700 laying hens in two attached Eggmobiles
48 nest boxes per Eggmobile

EVER SINCE JOEL SALATIN coined the phrase "eggmobile" in his book *Pastured Poultry Profit$*, it seems that every pastured-poultry producer has built a mobile coop for his or her laying hens. This is the original Eggmobile, times two! Polyface Farm's first mobile egg-production model consisted of one 12-foot × 20-foot henhouse built on a trailer. After this model proved successful for raising pastured laying hens, Joel built another Eggmobile and connected it to the first one. "The two units together house 700 birds (350 apiece), and we move them every day for much of the spring, and then every two days," says Joel.

The coops follow behind the cattle so the hens can range freely through the pastures. "The water comes from our gravity-fed livestock system through hoses that wind up on a garden-hose reel attached to the Eggmobile," Joel explains. "It holds 300 feet of ⅜-inch hose so we can reach out to the trunk line at the edge of the field." Some feed is stored in barrels attached to the front of one of the coops. More feed is stored in a large bin inside the other coop. The coop floors are wooden slats covered with chicken wire. Nesting boxes line one wall and are accessed from the outside. A chicken door with chicken ramp is located at the base of each coop. Predator protection is provided at night by closing the coop doors. Joel believes that frequently moving the Eggmobile keeps predators confused and at bay during the day.

This model works well for small-scale egg production. The biggest drawback is that it requires about 50 acres of pasture. (Because the birds have a strong homing instinct, they need to be far away from familiar surroundings to ensure they lay at the Eggmobile.) For large-scale egg production, Joel prefers to use the hoop-house-skid model (see page 148).

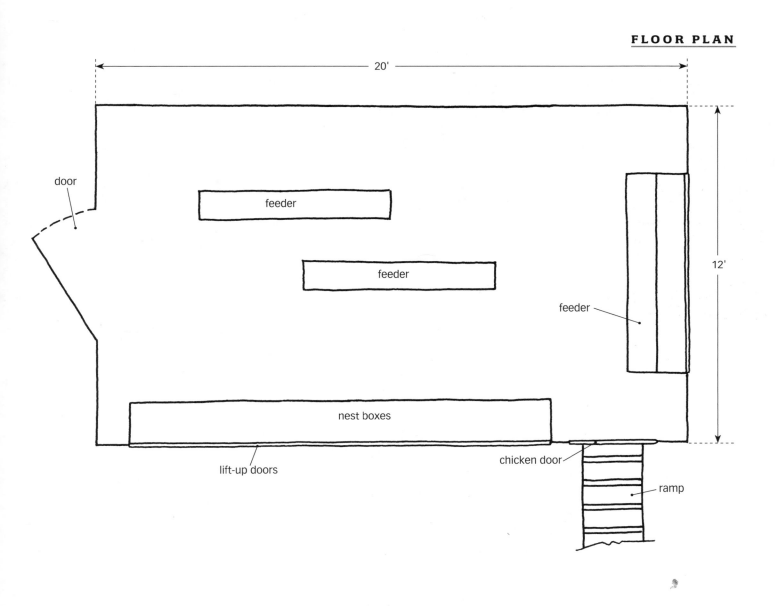

door

20'

feeder

feeder

feeder

12'

nest boxes

lift-up doors

chicken door

ramp

Salatin Eggmobile (continued)

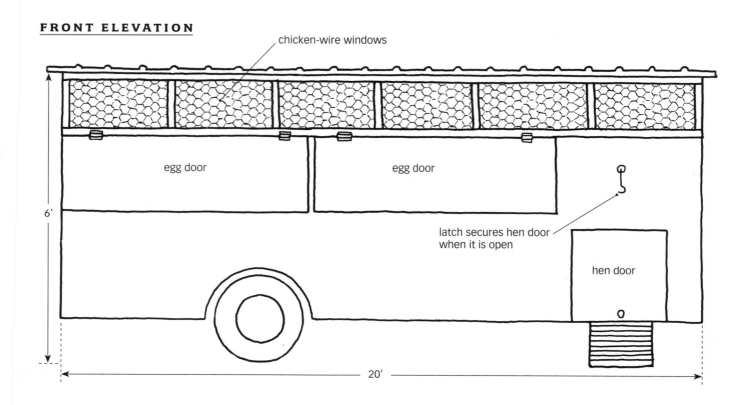

chicken-wire windows

egg door

egg door

latch secures hen door when it is open

hen door

6'

20'

HEN DOORS

Hen doors are openings cut into the side of the coop that allow hens access to the outdoors, as demonstrated by the Salatin Eggmobile and the Sap Bush Brooder House. Securing the hen door at night after the chickens have gone into the coop and are roosting provides predator protection. If the coop is surrounded by adequate fencing such as electric netting, this step may not be necessary. An easy way to move a mobile coop is to close the hens in at night and move the coop early the next morning. If the coop sits high off the ground, the hens find a wooden chicken ladder very helpful.

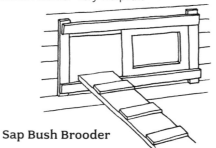

Sap Bush Brooder

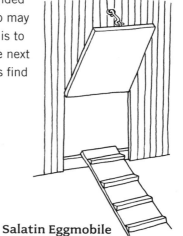

Salatin Eggmobile

Salatin Raken House

(see color photo on page 155)

HOLDS:
300 hens
48 nest boxes

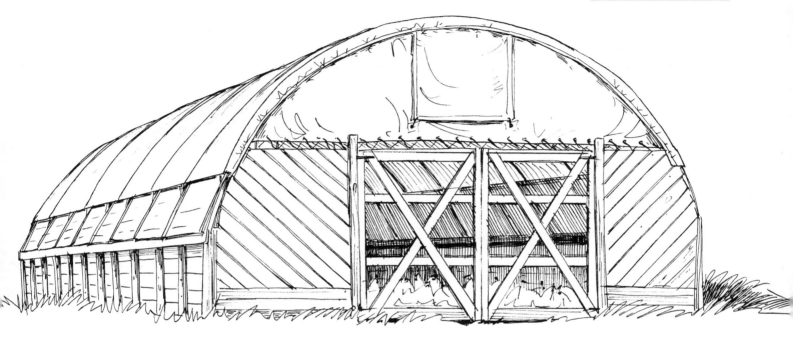

ON POLYFACE FARM, the Raken House is a combination henhouse and rabbit house originally built to house Daniel Salatin's meat-rabbit operation. The large stationary hoop house was purchased as a kit. Split locust posts act as the framework for the end walls and doors. Three-foot-long pipes were driven into the ground on the outside of the locust-post wall to act as columns and holders for the hoops. The insides of the sidewalls were covered with 1½-inch × 8-inch oak boards to hold in the bedding and to lock the fence posts together. Four locust poles were installed at the ends of the hoop house; two are 12 feet apart, and the other two are 6 feet beyond these. On the front end, siding was installed on the diagonal from the doors out to the edges of the house. The rear-end wall does not have doors, and its siding was also installed on the diagonal. The front and rear siding extends approximately 8 feet high until it meets the canvas hung on the top third of the ends.

The laying hens range about on the floor, and the rabbits are in cages at eye level along the wall. Nest boxes, built back-to-back on a tripod-type stand, are set in the center of the building. Feeders and waterers are placed on the floor. There are no roosts. Wood shavings are added every few days to reduce odors and provide a clean, dry, deep bedding pack. Occasionally Joel clears out the hens and runs a batch of young pigs through for a few weeks. The pigs — or, as Joel calls them, pigaerators — aerate the pack and create organic matter that is spread on the fields.

Salatin Raken House (continued)

FLOOR PLAN

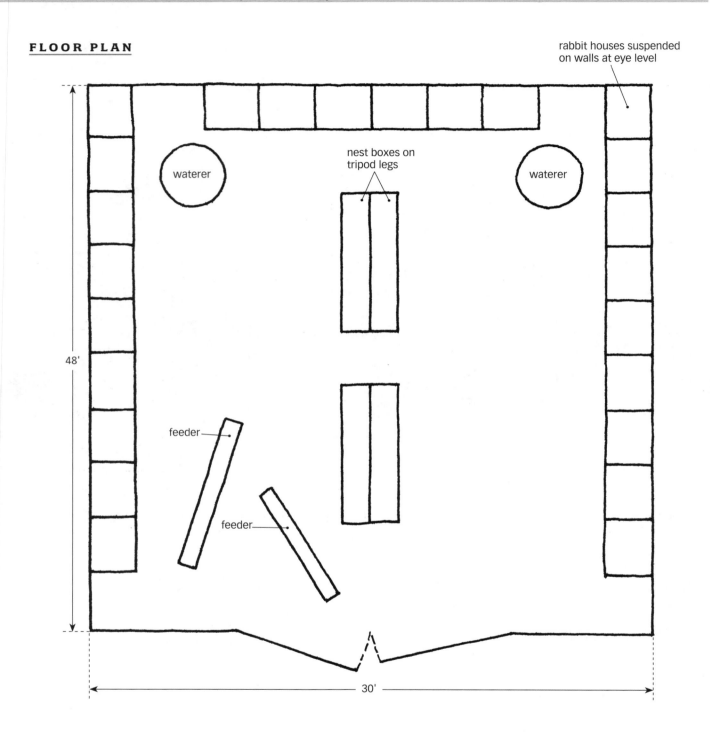

rabbit houses suspended on walls at eye level

nest boxes on tripod legs

waterer

waterer

feeder

feeder

48'

30'

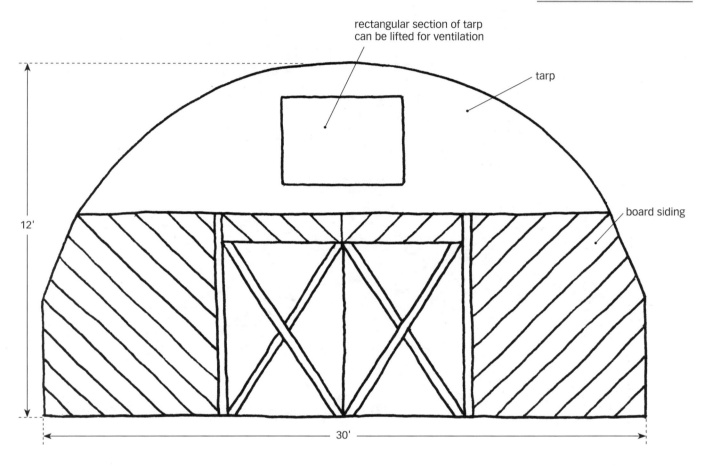

rectangular section of tarp
can be lifted for ventilation

tarp

board siding

12'

30'

Salatin Chick Brooder

JOEL SALATIN produces thousands of broilers and eggs from laying hens each year on Polyface Farm in Virginia's Shenandoah Valley. Chicks are started in a 20-foot × 60-foot brooder house (see page 1 for more information on chick brooders). When we visited, the last batch was in the brooder and brought the total for the season to 15,000 birds. That is a lot of chicken to produce and direct-market!

The wood-framed, shed-style barn, built specifically to be a chick brooder, is rat-proof with a concrete floor and metal siding. Wood shavings provide the bedding. Tall windows reach almost to the ground on all sides of the building and allow for ventilation in the summer. The Salatin Chick Brooder has three 20-foot × 20-foot divisions split by a chicken-wire wall to allow for separate batches of chicks. Metal trough feeders are placed on the floor, and height-adjustable automatic waterers are hung from the ceiling in each section. State-of-the-art industrial propane heaters, used per industrial specifications, do an excellent job of keeping the chicks warm. Broilers are raised in the brooder house for two to four weeks before being moved to the pasture coops.

CONSTRUCTION NOTES

▶ The building is framed with 2×4s and
has tin walls and a tin roof.

▶ Windows open out from the bottom
and hinge on the top.

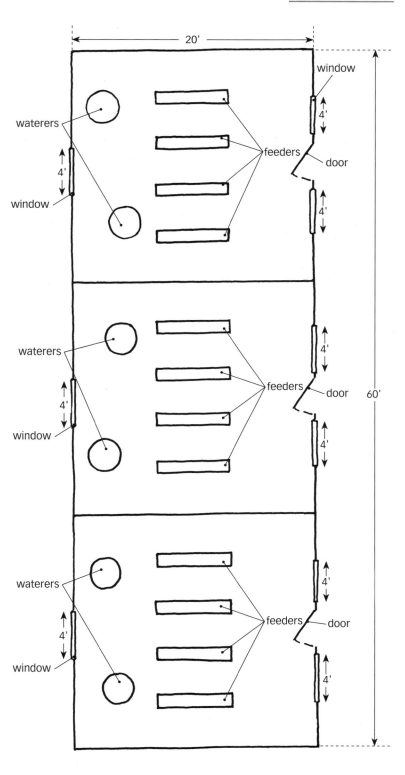

Winter Laying-Hen Hoop House

(see color photo on page 155)

HOLDS:
800 hens
80 nest boxes

AT POLYFACE FARM, three stationary hoop houses double as turkey brooders in the summer and housing for laying hens in the winter. Rabbit hutches also line one wall and rest above the birds. The two 20-foot × 120-foot and one 30-foot × 120-foot hoop houses were purchased as kits (see page 159 for company information).

To accommodate an 18-inch-deep bedding pack for the laying hens, sidewalls were built from split locust posts nailed to upright stakes. Three-foot-long pipes, or columns, were driven into the ground on the outside of the locust-post wall, and the hoops slip into the columns. The hoop-house covering is webbed, 9-mil poly that can be rolled up or down for ventilation. To protect the cover from damage, poultry netting is attached from a furring strip, fastened about 6 feet off the floor on the inside of the hoops, to the top of the stub walls. The bottom few feet of the hoop houses are lined with chicken wire to help keep the birds in and the predators out.

The hoop houses have two sets of double doors. The inside doors are made of poultry netting and hinge on the doorposts. The outside doors are made of corrugated metal and slide on a track. Both sets of doors are closed at night, but the outer set is opened during the day for ventilation. There are no roosts. The birds sleep on the floor.

When the chickens come out of the hoop house, Joel plants tomatoes and sweet corn inside, giving him a six-week jump start on the growing season.

CONSTRUCTION NOTES

▶ The hoop house is 12 feet high and was purchased as a kit.

▶ Sidewalls are built from split locust posts.

▶ Hoop-house covering is webbed 9-mil poly.

▶ The bottom 3 feet of the house is lined with chicken wire to keep out predators.

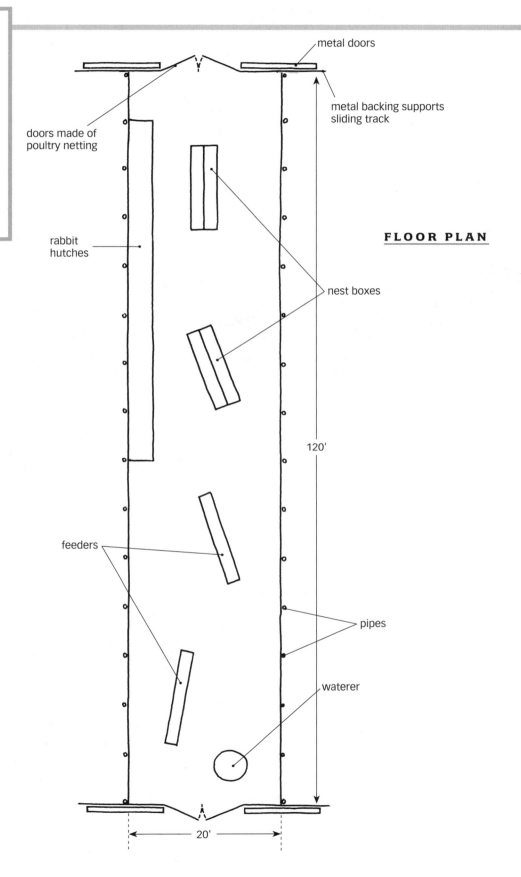

metal doors

metal backing supports sliding track

doors made of poultry netting

rabbit hutches

FLOOR PLAN

nest boxes

120'

feeders

pipes

waterer

20'

Salatin Feathernet System

(see color photo on page 155)

(see color photo on page 155)

HOLDS:
- 1,000 laying hens
- 72 nest boxes per hoop house

JOEL SALATIN of Polyface Farm developed the Feathernet System as a large-scale pastured-egg-production model. The Feathernet unit consists of two 20-foot × 20-foot hoop houses that shelter a total of one thousand birds. These coops are built from a kit (see page 159 for supplier information). The houses are secured to log skids that in turn are chained to skids holding bulk feeders with enough feed to last three weeks. Nest boxes are hung from the center of the hoops, and eggs are collected around 4 P.M. "At that time, we close up the perch boards in front of the nest boxes to keep the birds from being able to roost in there at night," Joel says.

Automatic waterers are connected to the farm's stock-water system. A circle of 450 feet of electrified poultry netting surrounds the hoop houses and keeps out predators. The coops are open on each end, giving the hens free access to the pastures within the netting. A second 450-foot fence is set up adjacent to the first. After three days, the entire skid system is pulled by a tractor to the new fenced area. The fence from the original section is then positioned in front so it is ready for the next move. The Feathernet System requires 5 acres of pasture. Joel grazes cattle on the pastures to keep the grass short for both the laying hens and the broilers.

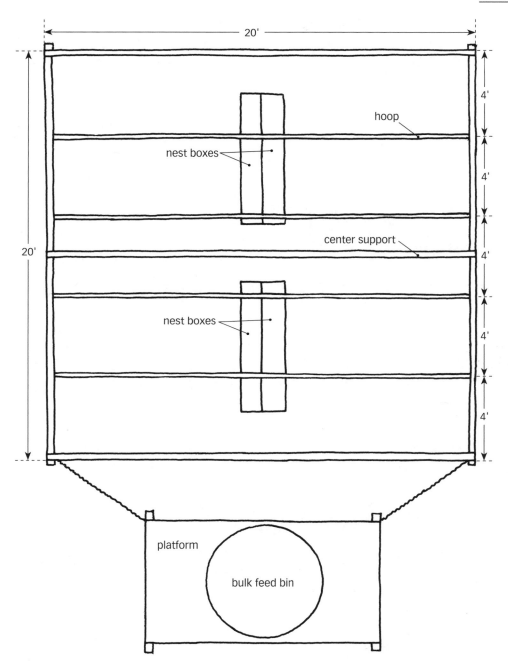

FRONT ELEVATION

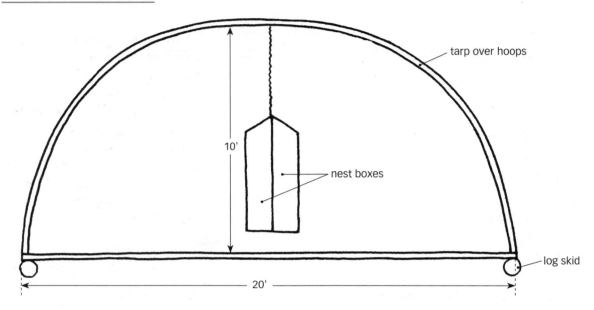

tarp over hoops

10'

nest boxes

log skid

20'

SALATIN MOBILE BULK FEED STORAGE

Joel Salatin uses old, cleaned fuel tanks to store feed for his pastured meat birds and laying hens. These bins are covered with tin roofing, mounted on log skids, and attached with chains to the hoop-house skids. When the hoop houses are moved, the feeders are stacked on top of the skid platforms and the entire system is moved all at once.

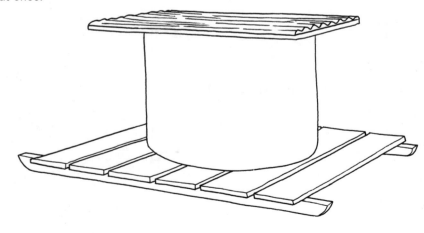

Cool Coops

The featured coop owners told me that their hens were aflutter with the hope that their coops would be featured in the photo gallery! Many of the coops (and hens, of course!) illustrated in the previous sections are depicted in the "Selected Coops" section. Coops that were not included in the book, but that provide lots of ideas and inspiration, are depicted in the "Inspiring Coops" section that follows.

SELECTED COOPS
(instructions found in this book)

CÉZANNE'S GARDEN COOP, PAGE 19

OAKHURST COMMUNITY GARDEN COOP,
PAGE 33

POULET CHALET, PAGE 47

CHICKS 'N THE 'HOOD COOP, PAGE 68

CORDWOOD CHICKEN HOUSE, PAGE 36

SHILOH MINI HEN RETREAT, PAGE 109

CHICKS 'N THE 'HOOD NEST BOX,
PAGE 68

CAT HOUSE, PAGE 51

JUDY PANGMAN

COOP DE GRASS, PAGE 126

DENNIS HARRISON-NOONAN

KIDS' GARDEN COOP, PAGE 65

MARK AND JODI CLAGG

© MARY LANGENFELD PHOTO

SHAKE YOUR TAIL FEATHERS COOP,
PAGE 55

PETER POIRIER

LITTLE RED HENHOUSE, PAGE 40

CRANE A. STAVIG

STARCLUCKS COOP, PAGE 61

JENNIFER CARLSON, LANDSCAPE DESIGNER

HI-RISE COOP, PAGE 44

JUDY PANGMAN

SUN COOP, PAGE 23

DEBRA COMPTON

SAN MIGUEL COOP, PAGE 12

SHARON ELY

TOOLSHED HENHOUSE, PAGE 57

JUDY PANGMAN

SAP BUSH EGGMOBILE, PAGE 92

SAP BUSH BROILER & TURKEY HOOPS,
PAGE 85

SALATIN FEATHERNET SYSTEM, PAGE 148

CHICKEN PRAIRIE SCHOONER, PAGE 81

SETTING HEN HUT, PAGE 10

SALATIN RAKEN HOUSE INTERIOR,
PAGE 141

SALATIN EGGMOBILE, PAGE 138

INSPIRING COOPS

with more ideas for construction and decoration

These unusual coops are shown to inspire your own unique design ideas.
(Plans for these coops are not available through Storey Publishing.)

This simple shed-style coop in Wisconsin provides for these chickens' basic needs.

A small greenhouse adjoins this chicken coop. Its owner, a carpenter, built the entire structure using recycled materials.

This stylish "eglu" is available from Omlet USA (see page 159 for contact information).

A combination of new and recycled materials was used to build this coop. In nice weather, the hens are allowed into the backyard. Because it is next to a popular bike path, "escapees" never go unnoticed.

Chicken-wire walls provide shelter for the birds at a heritage and rare breeds ranch in northern California.

Affectionately dubbed the Hensington Palace, this coop was made using a prefab garden shelter that was attached to an existing garage. The windows were purchased at a surplus store and added to it.

A converted rabbit hutch houses these chickens nicely.

Innkeeper Donna Murphy built this coop entirely out of recycled materials. It sits near her bed-and-breakfast in Wisconsin.

Using a German technique called fachwerk, the owners first constructed a timber framework and then filled panels with brick and mortar.

This charming little coop was made mostly of recycled material and leftover scraps from a house remodel.

This recreation of a nineteenth-century Southern coop was covered with hand-split wooden shingles. It is part of the Historic Latta Plantation in North Carolina.

An old headboard provides a bit of separation between the coop and the garden "bed" beside it.

Here's one way to put your eggs on display!

This cute coop was made from a kit available at Wine Country Coops (see page 159 for contact information).

This chicken "sits" outside the Karstens' coop in Wisconsin.

Sitting high atop an old telegraph pole, this creative English coop is far out of the way of foxes. The chickens climb the ladder to lay and to sleep at night.

While often left free to roam in the enclosed yard, the birds can find shelter in this roofed chicken run. The inside of this Midwestern coop was insulated with old window coverings.

Basement-type replacement windows provide ventilation and access to the chicken run. Hardware cloth was added to the top window to prevent access by predators.

Suppliers

Coops and Suppliers

Backyard Farming
Woodinville, Washington
206-715-1359
www.backyardfarming.com
"Stylish coops for suburban chicks."
United States distributor of products
made by Forsham Cottage Arks

Benedict Antique Lumber and Stone
Susquehanna, Pennsylvania
570-756-7878
www.benedictbarns.com
Garden sheds, children's playhouses,
and outhouses that can be easily
converted to chicken coops. Custom-
built from antique materials

The Carriage Shed
White River Junction, Vermont
800-441-6057
www.carriageshed.com
Amish-made chicken coops

Cold Springs Farm
Sharon Springs, New York
518-234-8320
www.coldspringsorganic.com
Variety of livestock feeds, including
feed used by Judy Pangman

Cutler's Pheasant and Poultry Supply, Inc.
Applegate, Michigan
810-633-9450
www.cutlersupply.com
Brooders and a variety of poultry
supplies

EggCartons.com
Manchaug, Massachusetts
888-852-5340
www.eggcartons.com
Egg cartons, egg trays, and poultry
supplies

Jamaica Cottage Shop, Inc.
Jamaica, Vermont
866-297-3760
www.jamaicacottageshop.com
Fine garden buildings built by
Vermont tradesmen

Omlet Ltd.
866-653-8872
www.omlet.us
High-end chicken houses

Wine Country Coops
Sebastopol, California
707-829-8405
www.winecountrycoops.com
Custom-designed chicken coops

Winkler Canvas Ltd.
Winkler, Manitoba
800-852-2638
www.winklercanvasbldg.com
Canvas structures, including hoop
house kits used by Joel Salatin

Hatcheries

Belt Hatchery
Fresno, California
559-264-2090
www.belthatchery.com
Chickens

Cackle Hatchery
Lebanon, Missouri
417-532-4581
www.cacklehatchery.com
Chickens, turkeys, ducks, geese,
and guineas

Double R Country Store
Palm Bay, Florida
321-837-1625
www.dblrsupply.com
Chickens, turkeys, ducks, geese,
and guineas

Dunlap Hatchery
Caldwell, Idaho
208-459-9088
www.dunlaphatchery.net
Chickens, turkeys, ducks, geese,
and guineas

Hoffman Hatchery, Inc.
Gratz, Pennsylvania
717-365-3694
www.hoffmanhatchery.com
Chickens, turkeys, ducks, geese,
and guineas

Hoover's Hatchery
Rudd, Iowa
800-247-7014
www.hoovershatchery.com
Chickens, turkeys, ducks, geese,
and guineas

Ideal Poultry Breeding Farms, Inc.
Cameron, Texas
254-697-6677
www.ideal-poultry.com
Chickens, turkeys, ducks, geese,
and guineas

Murray McMurray Hatchery
Webster City, Iowa
800-456-3280
www.mcmurrayhatchery.com
Chickens, turkeys, ducks, geese,
and guineas

Privett Hatchery, Inc.
Portales, New Mexico
877-774-8388
www.privetthatchery.com
Chickens, turkeys, ducks, geese,
and guineas

Ridgway Hatchery, Inc.
LaRue, Ohio
800-323-3825
www.ridgwayhatchery.com
Chickens, turkeys, ducks, geese,
and guineas

Sand Hill Preservation Center
Calamus, Iowa
563-246-2299
www.sandhillpreservation.com
Chickens, turkeys, ducks, geese,
and guineas

Schlecht Hatchery
Miles, Iowa
563-682-7865
www.schlechthatchery.com
Chickens, ducks, and turkeys

Stromberg's Chicks and Game Birds
Pine River, Minnesota
800-720-1134
www.strombergschickens.com
Chickens, turkeys, ducks, geese,
and guineas

Welp, Inc.
Bancroft, Iowa
800-458-4473
www.welphatchery.com
Chickens, turkeys, ducks, geese,
and guineas

Helpful Sources

Resources for Poultry Producers

Alternative Farming Systems Information Center
United States Department of Agriculture
301-504-6559
http://afsic.nal.usda.gov
Collects, organizes, and distributes
information on alternative agriculture
and provides high-level searching and
reference services from the National
Agricultural Library's vast collection and
world-wide databases.

American Pastured Poultry Producers Association
www.apppa.org
The APPPA was established in 1997 to
assist all pastured poultry producers
in North America. It provides
opportunities through conferences
and newsletters for people to learn and
exchange information about raising
poultry on pasture.

ATTRA — National Sustainable Agriculture Information Service
National Center for Appropriate
Technology
800-346-9140
https://attra.ncat.org
Provides information and other
technical assistance to farmers, ranchers,
Extension agents, educators, and others
involved in sustainable agriculture in the
United States. Its in-depth publications
include information on production
practices, alternative crop and livestock
enterprises, innovative marketing, and
organic certification.

Sustainable Agriculture Research and Education (SARE)
www.sare.org
A program of the USDA's National
Institute of Food and Agriculture
(NIFA), SARE funds projects and
conducts outreach designed to improve
agricultural systems. It also publishes
books, introductory bulletins, and
education guidelines.

Sources for Sustainably Raised Poultry and Eggs

Eatwild.com
www.eatwild.com
This website was created by Jo Robinson (author of Pasture Perfect and *The Omega Diet*, with Dr. Artemis Simopoulos) to be the "clearinghouse for information about pasture-based farming." It has the latest news and information about the benefits of raising animals on pasture (for the consumer, farmer, and animal) and has a free directory of sustainably raised meat, poultry, dairy, and eggs from stores, farms, and restaurants in your area.

Heritage Foods USA
718-389-0985
www.heritagefoodsusa.com
An independently owned company dedicated to saving Native American turkeys, pigs, sheep, bison, cows, reef-net salmon, chickens, and all breeds of food livestock. It sells heritage meat, fish, poultry, and grains.

Slow Food USA
877-756-9366
www.slowfoodusa.org
A nonprofit educational organization dedicated to supporting and celebrating the food traditions of North America. Local chapters advocate sustainability and biodiversity through educational events and public outreach. They also publish regional guides to Slow Food member restaurants.

Chicken-Related Web Resources

BackYard Chickens.com
www.backyardchickens.com
Information about chickens and coop designs

Cezanne's Garden
(formerly the website of Jana Barnhart and her chickens)

Chicken Resources on the Web
Ithaca College
www.ithaca.edu/staff/jhenderson/chooks/chlinks.html
Interesting list of chicken-related Web resources. Complied by John R. Henderson of Sage Hen Farm in Lodi, New York.

Free-Range Poultry
http://free-rangepoultry.com
Herman Beck-Chenoweth's website; includes a history of the free-range system and a sample budge for a free-change chicken operation

Isthmus Handyman LLC
www.isthmushandyman.com
Website of Dennis Harrison-Noonan. Provides information on the Kids' Garden Coop.

Mad City Chickens
www.madcitychickens.com
An organization in Madison, Wisconsin, that offers information and support to backyard chicken owners

Related Reading

Barnyard in Your Backyard, edited by Gail Damerow. Storey Publishing, 2002.

The Chicken Health Handbook, by Gail Damerow. Storey Publishing, 1994.

Family Friendly Farming: A Multi-Generational Home-Based Business Testament, by Joel Salatin. Polyface, 2001.

Free-Range Poultry Production and Marketing: A Guide to Raising, Processing and Marketing Premium Quality Chicken & Turkey & Eggs, by Herman Beck-Chenoweth. Back Forty Books, 1997.

The Grassfed Gourmet Cookbook, by Shannon Hayes. Ten Speed Press, 2005.

Holy Cows and Hog Heaven: The Food Buyer's Guide to Farm Friendly Food, by Joel Salatin. Polyface, 2005.

How to Build Animal Housing, by Carol Ekarius. Storey Publishing, 2004.

Keep Chickens!, by Barbara Kilarski. Storey Publishing, 2003.

Pastured Poultry Profit$, by Joel Salatin. Polyface, 1996.

Salad Bar Beef, by Joel Salatin. Polyface, 1996.

Small-Scale Livestock Farming, by Carol Ekarius. Storey Publishing, 1999.

Storey's Guide to Raising Chickens, by Gail Damerow. Storey Publishing, 1995.

Storey's Guide to Raising Poultry, by Leonard S. Mercia. Storey Publishing, 2001.

Storey's Guide to Raising Turkeys, by Leonard S. Marcia. Storey Publishing, 2001.

Storey's Illustrated Breed Guide to Sheep, Goats, Cattle, and Pigs, by Carol Ekarius. Storey Publishing, 2008.

You Can Farm: The Entrepreneur's Guide to Start & Succeed in a Farming Enterprise, by Joel Salatin. Polyface, 1998.

Index

Page numbers in *italics* indicate photographs. Page numbers in **boldface** indicate tables.